ECONOMIC MODELS

Methods, Theory and Applications

ECONOMIC MODELS

Methods, Theory and Applications

editor

Dipak Basu

Nagasaki University, Japan

World Scientific

NEW JERSEY • LONDON • SINGAPORE • BEIJING • SHANGHAI • HONG KONG • TAIPEI • CHENNAI

Published by

World Scientific Publishing Co. Pte. Ltd.
5 Toh Tuck Link, Singapore 596224
USA office: 27 Warren Street, Suite 401-402, Hackensack, NJ 07601
UK office: 57 Shelton Street, Covent Garden, London WC2H 9HE

British Library Cataloguing-in-Publication Data
A catalogue record for this book is available from the British Library.

ECONOMIC MODELS
Methods, Theory and Applications

ISBN-13 978-981-283-645-8
ISBN-10 981-283-645-4

Typeset by Stallion Press
Email: enquiries@stallionpress.com

Printed in Singapore by World Scientific Printers

This Volume is Dedicated to the Memory of
Prof. Tom Oskar Martin Kronsjo

Contents

Tom Oskar Martin Kronsjo: A Profile

Lydia Kronsjo

University of Birmingham, England

Tom Kronsjo was born in Stockholm, Sweden, on 16 August 1932, the only child of Elvira and Erik Kronsjo. Both his parents were gymnasium teachers.

In his school days Tom Kronsjo was a bright, popular student, often a leader among his school contemporaries and friends. During his school years, he organized and chaired a National Science Society that became very popular with his school colleagues.

From an early age, Tom Kronsjo's mother encouraged him to learn foreign languages. He put these studies to a good test during a 1948 summer vacation when he hitch-hiked through Denmark, Germany, the Netherlands, England, Scotland, Ireland, France, and Switzerland. The same year Tom Kronsjo founded and chaired a technical society in his school. In the summer vacation of 1949, Tom Kronsjo organized an expedition to North Africa by seven members of the technical society, their ages being 14–24. The young people traveled by coach through the austere post-Second World War Europe. All the equipment for travel, including the coach, were donated by sympathetic individuals and companies, who were impressed by an intelligent, enthusiastic group of youngsters, eager to learn about the world. This event caught a wide interest in the Swedish press at that time and was vividly reported.

Tom Kronsjo used to mention that as a particularly personally exciting experience of his school years he remembered volunteering to give a series of morning school sermons on the need for the world governmental organizations, which would encourage and support mutual understanding of the needs of the people to live in peace.

Tom Kronsjo's university years were spent at both sides of the great divide. Preparing himself for a career in international economics, he wanted to experience and see for himself the different nations' ways of life and points of view. He studied in London, Belgrade, Warsaw, Oslo, and Moscow, before returning to his home country, Sweden, to pass examinations and submit required work at the Universities of Uppsala, Stockholm and Lund for his Fil.kand. degree in economics, statistics, mathematics and Slavonic Languages, and Fil.lic. degree in economics. By then, Tom Kronsjo besides his mother tongue — Swedish, was fluent in six languages — German, English, French, Polish, Serbo-Croatian, and Russian. During this period, Tom was awarded a number of scholarships for study in Sweden and abroad.

Tom Kronsjo met his wife while studying in Moscow. He was a visiting graduate student at the Moscow State University and she a graduate student with the Academy of Sciences of the USSR. They married in Moscow in February 1962. A few years later, a very happy event in the family life was the arrival of their adoptive son from Korea in October 1973.

As a post-graduate student in economics, Tom Kronsjo was one of a new breed of Western social scientists that applied rigorous mathematical and computer-based modeling to economics. In 1963, Tom Kronsjo was invited by Professor Ragnar Frish, later Nobel prize winner in econometrics, to work under him and Dr. Salah Hamid in Egypt as a visiting Assistant Professor in Operations Research at the UAR Institute of National Planning. His pioneering papers on optimization of foreign trade were soon considered classic and brought invitations to work as a guest researcher by the GDR Ministry of Foreign Trade and Inner German Trade, and by the Polish Ministry of Foreign Trade.

In 1964, Tom Kronsjo was appointed an Associate Professor in Econometrics and Social statistics at the Faculty of Commerce and Social Sciences, the University of Birmingham, U.K. He and his wife moved to the United Kingdom.

In Birmingham, together with Professor R.W. Davies Tom Kronsjo developed diploma, and master of social sciences courses in national economic planning. The courses, offered from 1967, attracted many talented students from all over the world. Many of these students then stayed with Tom Kronsjo as research students and many of these people are today university professors, industrialists, and economic advisors. In 1969, at the age of 36, Tom Kronsjo was appointed Professor of Economic Planning. By then Tom Kronsjo had published over a hundred research papers.

Tom Kronsjo has always had a strong interest in educational innovations. He was among the first professors at the university to introduce a television-based teaching, making it possible to widen the scope of material taught.

Tom Kronsjo was generous towards his students with his research ideas, stimulating his students intellectually, unlocking wide opportunities for his students through this generosity. He had the ability to see the possibilities in any situation and taught his students to use them.

Tom Kronsjo's research interests were wide and multifaceted. One area of his interest was planning and management at various levels of the national economy. He developed important methods of decomposition of large economic systems, which were considered by his contemporaries a major contribution to the theory of mathematical planning and programming. These results were published in 1972 in a book "*Economics of Foreign Trade*" written jointly by Tom Kronsjo, Dr. Z. Zawada, and Professor J. Krynicki, both of the Warsaw University, and published in Poland.

In 1972, Tom Kronsjo was invited as a Visiting Professor of Economics and Industrial Administration at Purdue University, Purdue, USA.

In 1974, Tom Kronsjo was a Distinguished Visiting Professor of Economics, at San Diego State University, San Diego, USA.

In the year 1975, Tom Kronsjo found himself as a Senior Fellow in the Foreign Policy Studies Program of the Brookings Institution, Washington, D.C., USA. In collaboration with three other scientists, he was engaged in the project entitled "Trade and Employment Effects of the Multilateral Trade Negotiations". In the words of his colleagues, Tom Kronsjo made an indispensable contribution to the project. Tom Kronsjo's ingenious, tireless, conceptual, and implementing efforts made it possible to carry out calculations involving millions of pieces of information on trade and tariffs, permitting the Brookings model to become the most sophisticated and comprehensive model available at the time for investigation into effects of the current multilateral trade negotiations. In addition, Tom Kronsjo conceived and designed the means for implementing, perhaps the most imaginative portion of the project, the calculation of "optimal" tariff cutting formulas using linear programming and taking account of balance of payments and other constraints on trade liberalization. A monograph on "Trade, Welfare, and Employment Effects of Trade Negotiations in the Tokyo Round" by Dr. W.R. Cline, Dr. Noboru Kawanabe, Tom Kronsjo, and T. Williams was published in 1978.

Tom Kronsjo's general interests have been in the field of conflicts and co-operation, war and peace, ways out of the arms race, East-West joint enterprises and world development. He published papers on unemployment, work incentives, and nuclear disarmament.

Tom Kronsjo served as one of the five examiners for one of the Nobel prizes in economic sciences. From time to time, he was invited to nominate a candidate or candidates for the Nobel prize in economics.

Tom Kronsjo took early retirement from the University of Birmingham in 1988 and devoted subsequent years working with young researchers from China, Russia, Poland, Estonia, and other countries.

Tom Kronsjo was a great enthusiast of vegetarianism and veganism, of Buddhist philosophy, yoga, and inter-space communication, of aviation and flying. He learned to fly in the United States and held a pilot license since 1972.

Tom Kronsjo was a man who lived his life to the fullest, who constantly looked for ways to bring peace to the world. He was an exceptional individual who had the incredible vision and an ability to foresee new trends. He was a man of great charisma, one of those people of whom we say "he is a man larger than life".

Tom Oskar Martin Kronsjo died in June 2005 at the age of 72.

About the Editor

Prof. Dipak Basu is currently a Professor in International Economics in Nagasaki University in Japan. He has obtained his Ph.D. from the University of Birmingham, England and did his post-doctoral research in Cambridge University. He was previously a Lecturer in Oxford University — Institute of Agricultural Economics, Senior Economist in the Ministry of Foreign Affairs, Saudi Arabia and Senior Economist in charge of Middle East & Africa Division of Standard & Poor (Data Resources Inc). He has published six books and more than 60 research papers in leading academic journals.

Contributors

Athanasios Athanasenas is an Assistant Professor of Operations Research & Economic Development in the School of Administration & Economics, Institute of Technology and Education, Serres, Greece. He was educated in Colorado and Virginia Universities and received his Ph.D. from the University of Minnesota. He was a Full-Bright scholar.

Olav Bjerkholt, Professor in Economics, University of Oslo, was previously the Head of the Research at the Norwegian Statistical Office and an Assistant Professor of Energy Economics at the University of Oslo. He has obtained his Ph.D. from the University of Oslo. He was a Visiting Professor in the Massachusetts Institute of Technology and was a consultant to the United Nations and the IMF.

Christophe Deissenberg is a Professor in Economics, at Université de la Méditerranée and GREQAM, France. He is a co-editor of the *Journal of Optimization Theory and Applications and of Macroeconomic Dynamics*, a member of the editorial board of Computational Economics and of Computational Management Science. He is a member of the Advisory Board of the Society for Computational Economics and a member of the Management Committee of the ESF action COSTP10–Physics of Risk, and a team leader of the European STREP project EURACE.

Andrew Hughes Hallett is a Professor of Economics and Public Policy in the School of Public Policy at George Mason University, USA. Previously he was a Professor of Economics at Vanderbilt University, and the University of Strathclyde in Glasgow, Scotland. He was educated in the University of Warwick and London School of Economics, holds a doctorate from Oxford University. He was a Visiting Professor in Princeton University, Free University of Berlin, Universities of Warwick, Frankfurt, Rome, Paris X and in the Copenhagen Business School. He is a Fellow of the Royal Society of Edinburgh.

Fabrizio Iacone obtained a Ph.D. in Economics at the London School of Economics and is currently a Lecturer in the Department of Economics at the University of York, UK. He has already published several articles in leading journals.

Nikos Konidaris is currently a Ph.D. student at Hellenic Open University, School of Social Sciences, researching in the field of Business Administration. He holds an MA degree in Business and Management and a Bachelor's Degree in Economics & Marketing from Luton University UK.

Nikitas Spiros Koutsoukis is an Assistant Professor of Decision Modelling and Scientific Management in the Department of International Economic Relations and Development, Democritus University, Greece. He holds MSc and Ph.D. in decision modeling from Brunel University (UK). He is interested mainly in decision modeling and governance systems, with applications in risk management, business intelligence and operational research based decision support.

Alexis Lazaridis, Professor in Econometrics, Aristotle University of Thessaloniki, Greece, has obtained his Ph.D. from the University of Birmingham, England. He was a member of the Administrative Board of the Organization of Mediation and Arbitration, of the Administrative Board of the Centre of International and European Economic Law, and of the Supreme Labour Council of Greece. During his term of service as Head of the Economic Department Faculty of Law and Economics, Aristotle University of Thessaloniki, he was conferred an honorary doctoral degree by the President of the Republic of Cyprus. He was also pronounced an honorary member of the Centre of International and Economic European Law by the Secretary-General of the Legal Services of E.U. He has published several books, large number of scientific papers and has patents on two computer software programs.

Athanassios Mihiotis is an Assistant Professor at the School of Social Science of the Hellenic Open University. He has obtained his Ph.D. in industrial management from the National Technical University of Athens, Greece. He is the editor of the *International Journal of Decision Sciences, Risk and Management.* He has in the past worked as Planning and Logistics Director for multinational and domestic companies and has served as member of project teams in the Greek public sector.

Victoria Miroshnik, is currently an Adam Smith Research Scholar, Faculty of Law and Social Sciences, University of Glasgow, Scotland. She was educated in Moscow State University and was an Assistant Professor in Tbilisi State University, Georgia. She has published two books and a large number of scientific papers in leading management journals.

Anna-Maria Mouza, Assistant Professor in Business Administration, Technological Educational Institute of Serres, Greece, has obtained her Ph.D. from the School of Health Sciences, Aristotle University of Thessaloniki, Greece. Previously, she was an Assistant Professor in the School of Technology Applications of the Technological Institute of Thessaloniki, in the Department of Agriculture and in the Faculty of Health Sciences of the Aristotle University of Thessaloniki. She is the author of two books and many published scientific papers.

R. Orsi, Professor in Econometrics, University of Bologna, Italy, has obtained his Ph.D. from Universitè Catholique de Louvain, Belgium. He was previously educated in the University of Bologna and in the University of Rome. He was an Associate Professor in the University of Modena, and Professor of Econometrics in the University of Calabria, Italy. He was a Visiting Professor in the Center for Operation Research and Econometrics (CORE), Universitè-Catholique de Louvain, Belgium, Nato Senior Fellow, Head of the Department of Economics at the University of Calabria from 1987 to 1989, Director of CIDE (the Inter-Universitary Center of Econometrics) from 1997 to1999, and from 2000 to 2002, and the Head of Department of Economics, University of Bologna from 1999 to 2002.

Pavel Ševčík, was educated in the Université de la Méditerranée. He is presently completing his Ph.D in the Universilé de Montréal, Canada.

Introduction

We define "model building in economics," as the fruitful area of economics designed to solve real-world problems using all available methods available without distinctions: mathematical, computational, and analytical. Wherever needed we should not be shy to develop new techniques — whether mathematical or computational. This was the philosophy of Prof Tom Kronsjo, in whose memory we dedicate this volume. The idea is to develop stylized facts amenable to analytical methods to solve problems of the world.

The articles in this volume are divided into three distinct groups: methods, theory, and applications.

In the method section there are two parts: mathematical programming and computation of co-integrating vectors, which are widely used in econometric analysis. Both these parts are important to analyze and develop policies in any analytical model. Berjkholt has reviewed the history of development of mathematical programming and its impacts in economic modeling. In the process, he has drawn our attention to some methods, first developed by Ragner Frisch, which can provide solution where the standard Simplex method may fail. Frisch has developed these methods while working on the Second Five Year's Plan for India and has developed a method, which is similar to that of Karmarkar.

Lazaridis presents a completely new method to compute co-integration vectors, widely used in econometric analysis, by applying singular value decomposition. With this method, one can easily accommodate in the co-integrating vectors any deterministic factors, such as dummies, apart from the constant term and the trend. In addition, a comparatively simple procedure is developed for determining the order of integration of a different stationary series. Besides, with this procedure one can directly detect whether the differencing process produces a stationary series or not, since it seems to be a common belief that differencing a variable (one or more times) we will always get a stationary series, although this is not necessarily the case.

Basu and Lazaridis have proposed a new method of estimation and solution of an econometric model with parameters moving over time. This type of model is very realistic for economies, which are in transitional phases. The model was applied for the Indian economy to derive monetary-fiscal policy structure that would respond to the changing environments of the economy. The authors provide analysis to show that the response of the economy would be variable i.e., not at all fixed over time.

In the macroeconomic theory section Hughes-Hallet has examined the impacts of fiscal policy in a regime with independent monetary authority and how to co-ordinate public policies in such a regime. Independence of the central bank is a crucial issue for both developed and developing country. In this theoretical model, the author has discovered some important conclusions that fiscal leadership leads to improved outcomes because it implies a degree of co-ordination and reduced conflicts between institutions, without the central bank having to lose its ability to act independently.

Deissenberg and Pavel consider a dynamic model of environmental taxation that exhibits time inconsistency. There are two categories of firms, believers (who take the non-binding tax announcements made by the regulator at face value), and non-believers (who make rational but costly predictions of the true regulator's decisions). The proportion of believers and non-believers changes over time depending on the relative profits of both groups. If the regulator is sufficiently patient and if the firms react sufficiently fast to the profit differences, multiple equilibria can arise. Depending on the initial number of believers, the regulator will use cheap-talk together with actual taxes to steer the economy either to the standard, perfect prediction solution suggested in the literature, or to equilibrium with a mixed population of believers and non-believers who both make prediction errors. This model can be very useful in the analysis of environmental economics.

In the section on theory of business organization, Victoria Miroshnik has formulated a model of Japanese organization and how the superior corporate performances in the Japanese multi-national companies are the result of a multi-layer structure of cultures, national, personal, organizational. The model goes deep into the analysis of fundamental ingredient of Japanese organizational culture and human resource management system and how these contribute to corporate performance. The model produces a novel scheme, how to estimate phenomena which are seemingly unobserved psychological and sociological characteristics.

Mihiotis *et al.*, discussed enterprise modeling and integration challenges, and rationale as well as current tools and methodology for deeper analysis. Over the last decades, much has been written and a lot of research has been undertaken on the issue of enterprise modeling and integration. The purpose of this discussion is to provide an overview and understanding of the key concepts. The authors have outlined a framework for advancing enterprise integration modeling based on the state-of-the art techniques.

Anna Maria Mauza in a path breaking research presents a model suitable for an efficient budget management of a health service unit, by applying goal programming. She analyzes all the details needed to formulate the proper model, in order to successfully apply goal programming, in an attempt to satisfy the expectations of the decision maker in the best possible way, providing at the same time alternative scenarios considering various socio-economic factors.

In the section for policy analysis, Iacone and Orsi applied a small macro-econometric model to ascertain if the inflation dynamics and controls for Poland, Czech Republic, and Slovenia, are compatible with the remaining EU member countries. They found that the real exchange rate is the most effective instrument to stabilize inflation whereas direct inflation control mechanisms may be ineffective in certain cases. These experiments are very useful to design anti-inflation policies in open economies.

Athanasios Athanasenas investigated the co-integration dynamics of the credit–income nexus, within the economic growth process of the post-war US economy, over the period from 1957 up to 2007. Given the existing empirical research on the credit-lending channel and the established relationship between financial intermediation and economic growth in general, the main purpose is to analyze in detail the causal relationship between finance and growth by focusing on bank credit and income GNP, in the post-war US economy. This is a new application of an innovative technique of co-integration analysis with emphasis on system stability analysis. The results show that there is no short-run effect of credit changes on income changes, but only in the long-run, credit affects money income.

The book covers most of the important areas of economics with the basic analytical framework to formulate a logical structure and then suggest and implement methods to quantify the structure to derive applicable policies. We hope the book would be a source of joy for anyone interested to make economics a useful discipline to enhance human welfare rather than being a sterile discourse devoid of reality.

Chapter 1

Methods of Modelling: Mathematical Programming

Topic 1

Some Unresolved Problems of Mathematical Programming

Olav Bjerkholt
University of Oslo, Norway

1. Introduction

Linear programming (LP) emerged in the United States in the early post-war years. One may to a considerable degree see the development of LP as a direct result of the mobilization of research efforts during the war. George B. Dantzig, who was employed by the US Armed Forces, played a key role in developing the new tool by his discovery in 1947 of the Simplex method for solving LP problems. Linear programming was thus a military product, which soon appeared to have very widespread civilian applications. The US Armed Forces continued its support of Dantzig's LP work, as the most widespread textbook in LP in the 1960s, namely Dantzig (1963), was sponsored by the US Air Force.

Linear programming emerged at the same time as the first electronic computers were developed. The uses and improvements in LP as an optimization tool developed in step with the fast development of electronic computers.

A standard formulation of a LP problem in n variables is as follows:

$$\min_{x}\{c'x \,|\, Ax = b, x \geq 0\}$$

where $c \in R^n$, $b \in R^m$ and A a (m, n) matrix of rank $m < n$. The number of linear constraints is thus m. The feasibility region of the problem consists of all points fulfilling the constraints.

$$S = \{x \in R^n \,|\, Ax = b, x \geq 0\}$$

where S is assumed to be bounded with a non-empty interior.

Several leading economists, including a number of future Nobel Laureates, took active part in developing and utilizing LP at an early stage.

Among these was also Ragnar Frisch. Frisch was broadly oriented towards macroeconomic policy and planning problems and was highly interested in the promising new optimization tool. Frisch, who had a strong background in mathematics and also was very proficient in numerical methods, made the development of solution algorithms for LP problems a part of his research agenda. During the 1950s, Frisch invented and tested out various solution algorithms along other lines than the Simplex method for solving LP problems.

With regard to the magnitude of LP problems that could be solved at different times with the computer equipment at disposal, Orden (1993) gives the following rough indication. In the first year after 1950, the number m of constraints in a solvable problem was of order 100 and has grown with a factor of 10 for each decade, implying that currently LP problems may have a number of constraints that runs into tens of millions. Linear programming has been steadily taken into use for new kinds of problems and the computer development has allowed the problems to be bigger and bigger.

The Simplex method has throughout this period been the dominant algorithm for solving LP problems, not least in the textbooks. It is perhaps relatively rare that an algorithm developed at an early stage for new problem, as Dantzig did with the Simplex method for the LP problem, has retained such a position. The Simplex method will surely survive in textbook presentations and for its historical role, but for the solution of large-scale LP problems it is yielding ground to alternative algorithms. Ragnar Frisch's work on algorithms is interesting in this perspective and has been given little attention. It may on a modest scale be a case of a pioneering effort that was more insightful than considered at the time and thus did not get the attention it deserved. In the history of science, there are surely many such cases.

Let us first make a remark on the meaning of algorithms. The mathematical meaning of "algorithm" in relation to a given type of problem is a *procedure, which after finite number of steps finds the solution to the problem*, or determines that there is no solution. Such an algorithmic procedure can be executed by a computer program, an idea that goes back to Alan Turing. But when we talk about algorithms with regard to the LP problem it is not in the mathematical sense, but a more practical one. An LP algorithm is a *practically executable technique for finding the solution to the problem.* Dantzig's Simplex method is an algorithm in both meanings. But one may have an algorithm in the mathematical sense, which is not a practical algorithm (and indeed also *vice versa*).

In the mathematical sense of algorithm, the LP problem is trivial. It can easily be shown that the feasibility region is a convex set with a linear surface. The optimum point of a linear criterion over such a set must be in a corner or possibly in a linear manifold of dimension greater than one. As the number of corners is finite, the solution can be found, for example, by setting $n - m$ of the x's equal to zero and solve $Ax = b$ for the remaining ones. Then all the corners can be searched for the lowest value of the optimality criterion.

Dantzig (1984) gives a beautiful illustration of why such a search for optimality is not viable. He takes a classical assignment problem, the distribution of a given number of jobs among the same number of workers. Assume that 70 persons shall be assigned to 70 jobs and the return of each worker-job combination is known. There are thus 70! possible assignment combinations. Dantzig's comment about the possibility of looking at all these combinations runs as follows:

> "Now 70! is a big number, greater than 10^{100}. Suppose we had an IBM 370-168 available at the time of the big bang 15 billion years ago. Would it have been able to look at all the 70! combinations by the year 1981? No! Suppose instead it could examine 1 billion assignments per second? The answer is still no. Even if the earth were filled with such computers all working in parallel, the answer would still be no. If, however, there were 10^{50} earths or 10^{44} suns all filled with nano-second speed computers all programmed in parallel from the time of the big bang until sun grows cold, then perhaps the answer is yes." (Dantzig, 1984, p. 106)

In view of these enormous combinatorial possibilities, the Simplex method is a most impressive tool by making the just mentioned and even bigger problems practically solvable. The Simplex method is to search for the optimum on the surface of the feasible region, or, more precisely, in the corners of the feasible region, in such a way that the optimality criterion improves at each step.

A completely different strategy to search for the optimum is to search the interior of the feasible region and approach the optimal corner (or one of them) from within, so to speak. It may seem to speak for the advantage of the Simplex method that it searches an area where the solution is known to be.

A watershed in the history of LP took place in 1984 when Narendra Karmarkar's algorithm was presented at a conference (Karmarkar, 1984a) and published later the same year in a slightly revised version (Karmarkar, 1984b). Karmarkar's algorithm, which the *New York Times* found worthy as first page news as a great scientific breakthrough, is such an "interior" method, searching the interior of the feasibility area in the direction of the optimal point.

This idea had, however, been pursued by Ragnar Frisch. To the best of my knowledge this was first noted by Roger Koenker a few years ago:

> "But it is an interesting irony, illustrating the spasmodic progress of science, that the most fruitful practical formulation of the interior point revolution of Karmarkar (1984) can be traced back to a series of Oslo working papers by Ragnar Frisch in the early 1950s." (Koenker, 2000).

2. Linear Programming in Economics

Linear programming has a somewhat curious relationship with academic economics. Few would today consider LP as a part of economic science. But LP was, so to say, launched within economics, or even within econometrics, and given much attention by leading economists for about 10 years or so. After around 1960, the ties to economics were severed. Linear programming disappeared from the economics curriculum and lost the attention of academic economists. It belonged from then on to operations research and management science on one hand and to computer science on the other.

It is hardly possible to answer to give an exact date for when "LP" was born or first appeared. It originated at around the same time as game theory with which it shares some features, and also at the time of some applied problems such as the transportation problem and the diet problem. Linear programming has various roots and forerunners in economics and mathematics in attempts to deal with economic or other problems using linear mathematical techniques. One such forerunner, but only slightly, was Wassily Leontief's input-output analysis. Leontief developed his "closed model" in the 1930s in an attempt to give empirical content to Walrasian general equilibrium. Leontief's equilibrium approach was transformed in his cooperation with the US Bureau of Labor Statistics to the open input-output model, see Kohli (2001).

In the early post-war years, LP and input-output analysis seemed within economics to be two sides of the same coin. The two terms were, for a short term, even used interchangeably. The origin of LP could be set to 1947, which, as already mentioned, was when Dantzig developed the Simplex algorithm. If we state the question as to when "LP" was first used in its current meaning in the title of a paper presented at a scientific meeting, the answer is to the best of my knowledge 1948. At the Econometric Society meeting in Cleveland at the end of December

1948, Leonid Hurwicz presented a paper on LP and the theory of optimal behavior with the first paragraph providing a concise definition for economists:

> "The term *linear programming* is used to denote a problem of a type familiar to economists: maximization (or minimization) under restrictions. What distinguishes linear programming from other problems in this class is that both the function to be maximized and the restrictions (equalities or inequalities) are linear in the variables." (Hurwicz, 1949).

Hurwicz's paper appeared in a session on LP, which must have been the first ever with that title. One of the other two papers in the session was by Wood and Dantzig and discussed the problem of selecting the "best" method of accomplishing any given set of objectives within given restriction as a LP problem, using an airlift operation as an illustrative example. The general model was presented as an "elaboration of Leontief's input-output model."

At an earlier meeting of the Econometric Society in 1948, Dantzig had presented the general idea of LP in a symposium on game theory. Koopmans had, at the same meeting, presented his general activity analysis production model. The conference papers of both Wood and Dantzig appeared in *Econometrica* in 1949. Dantzig (1949) mentioned as examples of problems for which the new technique could be used, Stigler (1945), known in the literature as having introduced the "diet problem," and a paper by Koopmans on the "transportation problem," presented at an Econometric Society meeting in 1947 (Koopmans, 1948).

Dantzig (1949) stated the LP problem, not yet in standard format, but the solution technique was not discussed. The paper referred not only Leontief's input-output model but also to John von Neumann's work on economic equilibrium growth, i.e., to recent papers firmly within the realm of economics. Wood and Dantzig (1949) in the same issue stated an airlift problem.

In 1949, Cowles Commission and RAND jointly arranged a conference on LP. The conference meant a breakthrough for the new optimization technique and had prominent participation. At the conference were economists, Tjalling Koopmans, Paul Samuelson, Kenneth Arrow, Leonid Hurwicz, Robert Dorfman, Abba Lerner and Nicholas Georgescu-Roegen, mathematicians, Al Tucker, Harold Kuhn and David Gale, and several military researchers including Dantzig.

Dantzig had discovered the Simplex method but admitted many years later that he had not really realized how important this discovery was. Few people had a proper overview of linear models to place the new discovery in context, but one of the few was John von Neumann, at the time an authority on a wide range of problems form nuclear physics to the development of computers. Dantzig decided to consult him about his work on solution techniques for the LP problem:

> "I decided to consult with the "great" Johnny von Neumann to see what he could suggest in the way of solution techniques. He was considered by many as the leading mathematician in the world. On October 3, 1947 I visited him for the first time at the Institute for Advanced Study at Princeton. I remember trying to describe to von Neumann, as I would to an ordinary mortal, the Air Force problem. I began with the formulation of the linear programming model in terms of activities and items, etc. Von Neumann did something, which I believe was uncharacteristic of him. "Get to the point," he said impatiently. Having at times a somewhat low kindling point, I said to myself "O.K., if he wants a quicky, then that's what he'll get." In under one minute I slapped the geometric and the algebraic version of the problem on the blackboard. Von Neumann stood up and said "Oh that!" Then for the next hour and a half, he proceeded to give me a lecture on the mathematical theory of linear programs." (Dantzig, 1984).

Von Neumann could immediately recognize the core issue as he saw the similarity with the game theory. The meeting became the first time Dantzig heard about duality and Farkas' lemma.

One could underline how LP once seemed firmly anchored in economics by pointing to the number of Nobel Laureates in economics who undertook studies involving LP. These comprise Samuelson and Solow, co-authors with Dorfman of Dorfman, Samuelson and Solow (1958), Leontief who applied LP in his input-output models, and Koopmans and Stigler, originators of the transportation problem and the diet problem, respectively. One also has to include Frisch, as we shall look at in more detail below, Arrow, co-author of Arrow *et al.* (1958), Kantorovich, who is known to have formulated the LP problem already in 1939, Modigliani, and Simon, who both were co-authors of Holt *et al.* (1956). Finally, to be included are also all the three Nobel Laureates in economics for 1990: Markowitz *et al.* (1957), Charnes *et al.* (1959) and Sharpe (1967). Koopmans and Kantorovich shared the Nobel Prize in economics for 1975 for their work in LP. John von Neumann died long before the Nobel Prize in economics was established, and George Dantzig never got the prize he ought to have shared with Koopmans and Kantorovich in 1975 for reasons that are not easy to understand.

Among all these Nobel Laureates, it seems that Ragnar Frisch had the deepest involvement with LP. But he published poorly and his achievements went largely unrecognized.

3. Frisch and Programming, Pre-War Attempts

Ragnar Frisch had a strong mathematical background and a passion for numerical analysis. He had been a co-founder of the Econometric Society in 1930 and to him econometrics meant both the formulation of economic theory in a precise way by means of mathematics *and* the development of methods for confronting theory with empirical data. He wrote both of these aims into the constitution of the Econometric Society as rendered for many years in every issue of *Econometrica*. He became editor of *Econometrica* from its first issue in 1933 and held this position for more than 20 years.

His impressive work in several fields of econometrics must be left uncommented here. Frisch pioneered the use of matrix algebra in econometrics in 1929 and had introduced many other innovations in the use of mathematics for economic analysis.

Frisch's most active and creative as a mathematical economist coincided with the Great Depression, which he observed at close range both in the United States and in Norway. In 1934, he published an article in *Econometrica* (Frisch, 1934a) where he discussed the organization of national exchange organization that could take over the need for trade when markets had collapsed due to the depression, see Bjerkholt and Knell (2006). In many ways, the article may seem naïve with regard to the subject matter, but it had a very sophisticated mathematical underpinning. The article was 93 pages long, the longest regular article ever published in *Econometrica.* Frisch had also published the article without letting it undergo proper referee process. Perhaps it was the urgency of getting it circulated that caused him to do this mistake for which he was rebuked by some of his fellow econometricians. The article was written as literally squeezed in between Frisch two most famous econometric contributions in the 1930s, his business cycle theory (Frisch, 1933) and his confluence theory (Frisch, 1934b).

Frisch (1934a) developed a linear structure representing inputs for production in different industries, rather similar to the input-output model of Leontief, although the underlying idea and motivation was quite different. Frisch then addressed the question of how the input needs could be modified in the least costly way in order to achieve a higher total production capacity of the entire economy. The variables in his problem (x_i) represented (relative) deviations from observed input coefficients. He formulated a cost function as a quadratic function of these deviations. His reasoning thus led to a quadratic target function to be minimized subject to a set of linear constraints. The problem was stated as follows:

$$\min \Omega = \frac{1}{2}\left[\frac{x_1^2}{\varepsilon_1} + \frac{x_2^2}{\varepsilon_2} + \cdots + \frac{x_n^2}{\varepsilon_n}\right] \quad \text{wrt. } x_k \quad k = 1, 2, \ldots, n$$

$$\text{subject to: } \Sigma_k f_{ik} x_k \geq C^* - \frac{a_{i0}}{P_i} \quad i = 1, \ldots, n$$

where C^* is the set target of overall production and the constraint the condition that the deviation allowed this target value to be achieved.

By solving for alternative values of C^*, the function $\Omega = \Omega(C^*)$, representing the cost of achieving the production level C^* could be mapped. Frisch naturally did not possess an algorithm for this problem. This did not prevent him from working out a numerical solution of an illuminating example and making various observations on the nature of such optimising problems. His solution of the problem for a given example with $n = 3$, determining Ω as a function of C^* ran over 12 pages in *Econometrica*, see Bjerkholt (2006).

Another problem Frisch worked on before the war with connections to LP was the diet problem. One of Frisch's students worked on a dissertation on health and nutrition, drawing on various investigations of the nutritional value of alternative diets. Frisch supervised the dissertation in the years immediately before World War II when it led him to formulate the diet problem, years before Stigler (1945). Frisch's work was published as the introduction to his student's monograph (Frisch, 1941). In a footnote, he gave a somewhat rudimentary LP formulation of the diet problem, cf. Sandmo (1993).

4. Frisch and Linear Programming

Frisch was not present at the Econometric Society meetings in 1948 mentioned above; neither did he participate in the Cowles Commission-RAND conference in 1949. He could still survey and follow these developments rather closely. He was still the editor of *Econometrica* and thus had accepted and surely studied the two papers by Dantzig in 1949. He had contact with Cowles Commission in Chicago, its two successive research directors in these years, Jacob Marschak and Tjalling Koopmans, and other leading members of the Econometric Society. Frisch visited in the early post-war years frequently in the United States, primarily due to his position as chairman of the UN Sub-Commission for Employment and Economic Stability. The new developments over the whole range of related areas of linear techniques had surely caught his interest.

He first touched upon LP in lectures to his students at the University of Oslo in September 1950. Two students drafted lecture notes that Frisch corroborated and approved. In these early lectures Frisch used "input-output analysis" and "LP" as almost synonymous concepts. The content of these early lectures was input-output analysis with emphasis on optimisation by means of LP, e.g., optimisation under a given import constraint or given labor supply, they did not really address LP techniques as a separate issue. Still these lectures may well have been among the first applying LP to the macroeconomic problems of the day in Western Europe, years before Dorfman *et al.* (1958) popularised this tool for economists' benefit. One major reason for Frisch pioneering efforts here was, of course, that the new ideas touched or even overlapped with his own pre-war attempts as discussed above. In the very first post-war years, he had indeed applied the ideas of Frisch (1934a) to the problem of achieving the largest possible multilateral balanced trade in the highly constrained world economy (Frisch, 1947; 1948).

The first trace of LP as a topic in its own right on Frisch's agenda was a single lecture he gave to his students on February 15th 1952. Leif Johansen, who was Frisch's assistant at the time, took notes that were issued in the Memorandum series two days later (Johansen, 1952). The lecture gave a brief mathematical statement of the LP problem and then as an important example discussed the problem related to producing different qualities of airplane fuel from crude oil derivates, with reference to an article by Charnes *et al.* in *Econometrica*. Charnes *et al.* (1952) is, indeed, a famous contribution in the LP literature as the first published paper on an industrial application of LP. A special thing about the Frisch's use of this example was, however, that his lecture did not refer the students to an article that had already appeared in *Econometrica*, but to one that was forthcoming! Frisch had apparently been so eager to present these ideas that he had simply used (or perhaps misused) his editorial prerogative in lecturing and issuing lecture notes on an article not yet published!

After the singular lecture on LP in February 1952, Frisch did not lecture on this topic till the autumn term in 1953, as part of another lecture series on input-output analysis. In these lectures, he went by way of introduction through a number of concrete examples of problem that were amenable to be solved by LP, most of which have already been mentioned above. When he came to the diet problem, it was not with reference to Stigler (1945), but to Frisch (1941). He exemplified, not least, with reconstruction in Norway, the building of dwellings under the tight post-war constraints and various macroeconomic problems.

In his brief introduction to LP, Frisch in a splendid pedagogical way used examples, amply illustrated with illuminating figures, and appealed to geometric intuition. He started out with cost minimization of a potato and herring diet with minimum requirements of fat, proteins and vitamin B, and then moved to macroeconomic problems within an input-output framework, somewhat similar in tenor to Dorfman (1958), still in the offing.

Frisch confined, however, his discussion to the formulation of the LP problem and the characteristics of the solution, he did not enter into solution techniques. But clearly he had already put considerable work into this. Because after mentioning Dantzig's Simplex method, referring to Dantzig (1951), Dorfman (1951) and Charnes *et al.* (1953), he added that there were two other methods, the *logarithm potential* method and the *basis variable* method, both developed at the Institute of Economics. He added the remark that when the number of constraints was moderate and thus the number of degrees of freedom large, the Simplex method might well be the most efficient one. But in other cases, the alternative methods would be more efficient. Needless to say, there were no other sources, which referred to these methods at this time. Frisch's research work on LP methods and alternatives to the Simplex method, can be followed almost step by step thanks to his working style of incorporating new results in the Institute Memorandum series. Between the middle of October 1953 and May 1954, he issued 10 memoranda. They were all in Norwegian, but Frisch was clearly preparing for presenting his work. As were to be expected Frisch's work on LP was sprinkled with new terms, some of them adapted from earlier works. Some of the memoranda comprised comprehensive numerical experiments. The memoranda were the following:

18 October 1953	*Logaritmepotensialmetoden til løsning av lineære programmeringsproblemer* [The logarithm potential method for solving LP problems], 11 pages.
7 November 1953	*Litt analytisk geometri i flere dimensjoner* [Some multi-dimensional analytic geometry], 11 pages.

(*Continued*)

13 November 1953	*Substitumalutforming av logaritmepotensialmetoden til løsning av lineære programmeringsproblemer* [Substitumal formulation of the logarithmic potential method for solving LP problems], 3 pages.
14 January 1954	*Notater i forbindelse med logaritmepotensialmetoden til løsning av lineære programmeringsproblemer* [Notes in connection with the logarithmic potential method for solving LP problems], 48 pages.
7 March 1954	*Finalhopp og substitumalfølging ved lineær programmering* [Final jumps and moving along the substitumal in LP], 8 pages.
29 March 1954	*Trunkering som forberedelse til finalhopp ved lineær programmering* [Truncation as preparation for a final jump in LP], 6 pages.
27 April 1954	*Et 22-variables eksempel på anvendelse av logaritmepotensialmetoden til løsning av lineære programmeringsproblemer* [A 22 variable example in application of the logarithmic potential method for the solution of LP problems].
1 May 1954	*Basisvariabel-metoden til løsning av lineære programmeringsproblemer* [The basis-variable method for solving LP problems], 21 pages.
5 May 1954	*Simplex-metoden til løsning av lineære programmeringsproblemer* [The Simplex method for solving LP problems], 14 pages.
7 May 1954	*Generelle merknader om løsningsstrukturen ved det lineære programmeringsproblem* [Generfal remarks on the solution structure of the LP problem], 12 pages.

At this stage, Frisch was ready to present his ideas internationally. This happened first at a conference in Varenna in July 1954, subsequently issued as a memorandum. During the summer, Frisch added another couple of memoranda. Then he went to India, invited by Pandit Nehru, and spent the entire academic year 1954–55 at the Indian Statistical Institute, directed by P. C. Mahalanobis, to take part in the preparation of the new five-year plan for India. While he was away, he sent home the MSS for three more memoranda. These six papers were the following:

June 21st 1954	*Methods of solving LP problems*, Synopsis of lecture to be given at the International Seminar on Input-Output Analysis, Varenna (Lake Como), June–July 1954, 91 pages.
August 13th 1954	*Nye notater i forbindelse med logaritmepotensialmetoden til løsning av lineære programmeringsproblemer* [New notes in connection with the logarithmic potential method for solving LP problems], 74 pages.
August 23rd 1954	*Merknad om formuleringen av det lineære programmeringsproblem* [A remark on the formulation of the LP problem], 3 pages.
October 18th 1954	*Principles of LP. With particular reference to the double gradient form of the logarithmic potential method*, 219 pages.
March 29th 1955	*A labour saving method of performing freedom truncations in international trade. Part I*, 21 pages.
April 15th 1955	*A labour saving method of performing freedom truncations in international trade. Part II*, 8 pages.

Before he left for India he had already accepted an invitation to present his ideas on programming in Paris. This he did on three different occasions in May–June 1955. He presented in French, but issued synopsis in English in the memorandum series. Thus, his Paris presentation were the following:

7th May 1955	*The logarithmic potential method for solving linear programming problems*, 16 pp. Synopsis of an exposition to be made on 1 June 1955 in the seminar of Professor René Roy, Paris (published as Frisch, 1956b).
13 May 1955	*The logarithmic potential method of convex programming. With particular application to the dynamics of planning for national developments*, 35 pp. Synopsis of a presentation to be made at the international colloquium in Paris, 23–28 May 1955 (published as Frisch, 1956c).
25 May 1955	*Linear and convex programming problems studied by means of the double gradient form of the logarithmic potential method*, 16pp. Synopsis of a presentation to be given in the seminar of Professor Allais, Paris, 26 May 1955.

After this, Frisch published some additional memoranda on LP. He launched a new method, named the Multiplex method, towards the end of 1955, later published in a long article in *Sankhya* (Frisch, 1955; 1957). Worth mentioning is also his contribution to the festschrift for Erik Lindahl in which he discussed the logarithmic potential method and linked it to the general problem of macroeconomic planning (Frisch, 1956a; 1956d).

From this time, Frisch's efforts subsided with regard to pure LP problems, as he concentrated on non-linear and more complex optimisation, drawing naturally on the methods he had developed for LP.

As Frisch managed to publish his results only to a limited degree and not very prominently there are few traces of Frisch in the international literature. Dantzig (1963) has, however, references to Frisch's work. Frisch also corresponded with Dantzig, Charnes, von Neumann and others about LP methods. Extracts from such correspondence are quoted in some of the memoranda.

5. Frisch's Radar: An Anticipation of Karmarkar's Result

In 1972, severe doubts were created about the ability of the Simplex method to solve even larger problems. Klee and Minty (1972), whom we must assume knew very well how the Simplex algorithm worked, constructed a LP problem, which when attacked by the Simplex method turned out to need a number of elementary operations which grew exponentially in the number of variables in the problem, i.e., the algorithm had exponential complexity. Exponential complexity in the algorithm is for obvious reasons a rather ruining property for attacking large problems.

In practice, however, the Simplex method did not seem to confront such problems, but the demonstration of Klee and Minty (1972) showed a worst case complexity, implying that it could never be proved what had been largely assumed, namely that the Simplex method only had polynomial complexity.

A response to Klee and Minty's result came in 1978 by the Russian mathematician Khachiyan 1978 (published in English in 1979 and 1980). He used a so-called "ellipsoidal" method for solving the LP problem, a method developed in the 1970s for solving convex optimisation problems by the Russian mathematicians Shor and Yudin and Nemirovskii. Khachiyan managed to prove that this method could indeed solve LP problems and had polynomial complexity as the time needed to reach a solution increased with the forth power of the size of the problem. But the method itself was hopelessly ineffective compared to the Simplex method and was in practice never used for LP problems.

A few years later, Karmarkar presented his algorithm (Karmarkar, 1984a; b). It has polynomial complexity, not only of a lower degree than Khachiyan's method, but is asserted to be more effective than the Simplex method. The news about Karmarkar's path-breaking result became front-page news in the *New York Times*. For a while, there was some confusion as to whether Karmarkar's method really was more effective than the Simplex method, partly because Karmarkar had used a somewhat special formulation of the LP problem. But it was soon confirmed that for sufficently large problems, the Simplex method was less effcient than Karmarkar's result. Karmarkar's contribution initiated a hectic new development towards even better methods.

Karmarkar's approach may be said to be to consider the LP just as another convex optimalization problem rather that exploiting the fact that the solution must be found in a corner by searching only corner points. One of those who has improved Karmarkar's method, Clovis C. Gonzaga, and for a while was in the lead with regard to possessing the nest method wrt. polynomial complexity, said the following about Karmarkar's approach:

> "Karmarkar's algorithm performs well by avoiding the boundary of the feasible set. And it does this with the help of a classical resource, first used in optimization by Frisch in 1955: the logarithmic barrier function:
>
> $$x \in R^n, \quad x > 0 \rightarrow p(x) = -\sum \log x_i.$$
>
> This function grows indefinitely near the boundary of the feasible set S, and can be used as a penalty attached to these points. Combining $p(\cdot)$ with the objective makes points near the boundary expensive, and forces any minimization algorithm to avoid them." (Gonzaga, 1992)

Gonzaga's reference to Frisch was surprising; it was to the not very accessible memorandum version in English of one of the three Paris papers.[1]

Frisch had indeed introduced the logarithmic barrier function in his logarithmic potential method. Frisch's approach was summarized by Koenker as follows:

> "The basic idea of Frisch (1956b) was to replace the linear inequality constraints of the LP, by what he called a log barrier, or potential function. Thus, in place of the canonical linear program

[1] In Gonzaga's reference, the author's name was given as K.R. Frisch, which can also be found in other rerences to Frisch in recent literature, suggesting that these refrences had a common source.

(1) $\min\{c'x|Ax = b, x \geq 0\}$

we may associate the logarithmic barrier reformulation

(2) $\min\{B(x, \mu)|Ax = b\}$

where

(3) $B(x, \mu) = c'x - \mu \sum \log x_k$

In effect, (2) replaces the inequality constraints in (1) by the penalty term of the log barrier. Solving (2) with a sequence of parameters μ such that $\mu \rightarrow 0$ we obtain in the limit a solution to the original problem (1)." (Koenker, 2000, p. 20)

Frisch had not provided exact proofs and he had above all not published properly. But he had the exact same idea as Karmarkar came up with almost 30 years later. Why did he not make his results better known? In fact there are other, even more important cases, in Frisch's work of not publishing. The reasons for this might be that his investigations had not yet come to an end. In the programming work Frisch pushed on to more complex programming problems, spurred by the possibility of using first- and second-generation computers. They might have seemed powerful to him at that time, but they hardly had the capacity to match Frisch's ambitions. Another reason for his results remaining largely unknown, was perhaps that he was too much ahead. The Simplex method had not really been challenged yet, by large enough problems to necessitate better method. The problem may have seen, not as much as a question of efficient algorithms as that of powerful enough computers.

We finish off with Frisch's nice illustrative description of his method in the only publication he got properly published (but unfortunately not very well distributed) on the logarithmic potential method:

> "Ma méthode d'approche est d'une espèce toute différente. Dans cette méthode nous travaillons systématiquement à l'intérieur de la région admissible et utilisons un potentiel logarithmique comme un guide– une sorte du radar–pour nous éviter de traverser la limite." (Frisch, 1956b, p. 13)

The corresponding quote in the memorandum synoptic note is:

> "My method of approach is of an entirely different sort [than the Simplex method]. In this method we work systematically in the *interior* of the admissible region and use a logarithmic potential as a guiding device — a sort of radar — to prevent us from hitting the boundary." (Memorandum of 7 May 1955, p. 8)

References

Arrow, KJ, L Hurwicz and H Uzawa (1958). *Studies in Linear and Non-Linear Programming*. Stanford, CA: Stanford University Press.

Bjerkholt, O and M Knell (2006). Ragnar Frisch and input-output analysis. *Economic Systems Research* **18**, (forthcoming).

Charnes, A, WW Cooper and A Henderson (1953). *An Introduction to Linear Programming*. New York and London: John Wiley & Sons.

Charnes, A, WW Cooper and B Mellon (1952). Bledning aviation gasolines — a study in programming interdependent activities in an integrated oil company, *Econometrica* **20**, 135–159.

Charnes, A, WW Cooper and MW Miller (1959). An application of linear programming to financial budgeting and the costing of funds. *The Journal of Business* **20**, 20–46.

Dantzig, GB (1949). Programming of interdependent activities: II mathematical model. *Econometrica* **17**, 200–211.

Dantzig, GB (1951). Maximization of a linear function of variables subject to linear inequalities. In *Activity Analysis of Production and Allocation*, T Koopmans (ed.), New York and London: John Wiley & Sons, pp. 339–347.

Dantzig, GB (1963). *Linear Programming and Extensions*. Princeton, NJ: Princeton University Press.

Dantzig, GB (1984). Reminiscences about the origin of linear programming. In *Mathematical Programming*, RW Cottle, ML Kelmanson and B Korte (eds.), Amsterdam: Elsevier (North-Holland), pp. 217–226.

Dorfman, R (1951). *An Application of Linear Programming to the Theory of the Firm*. Berkeley: University of California Press.

Dorfman, R, PA Samuelson and RM Solow, (1958). *Linear Programming and Economic Analysis*. New York: McGraw-Hill.

Frisch, R (1933). Propagation problems and impulse problems in economics. In *Economic Essays in Honor of Gustav Cassel*, R Frisch (ed.), London: Allen & Unwin, pp. 171–205.

Frisch, R (1934a). Circulation planning: proposal for a national organization of a commodity and service exchange. *Econometrica* 2, 258–336 and 422–435.

Frisch, R (1934b). *Statistical Confluence Analysis by Means of Complete Regression Systems*. Publikasjon nr 5, Oslo: Institute of Economics.

Frisch, R (1941). Innledning. In *Kosthold og levestandard. En økonomisk undersøkelse*, K Getz Wold (ed.), Oslo: Fabritius og Sønners Forlag, pp. 1–23.

Frisch, R (1947). On the need for forecasting a multilateral balance of payments. *The American Economic Review* **37**(1), 535–551.

Frisch, R (1948). The problem of multicompensatory trade. Outline of a system of multicompensatory trade. *The Review of Economics and Statistics* **30**, 265–271.

Frisch, R (1955). The multiplex method for linear programming. Memorandum from Institute of Economics, Oslo: University of Oslo (17 October 1955).

Frisch, R (1956a). Macroeconomics and linear programming. Memorandum from Institute of Economics, Oslo: University of Oslo, (10 January 1956). (Also published shortened and simplified as Frisch (1956d)).

Frisch, R (1956b). La résolution des problèmes de programme linéaire par la méthode du potential logarithmique. In *Cahiers du Séminaire d'Econometrie: No 4 — Programme*

Linéaire — Agrégation et Nombre Indices, pp. 7–20. (The publication also gives an extract of the oral discussion in René Roy's seminar, pp. 20–23.)

Frisch, R (1956c). Programme convexe et planification pour le développement national. *Colloques Internationaux du Centre National de la Recherche Scientifique LXII.* Les modèles dynamiques. Conférence à Paris, mai 1955, 47–80.

Frisch, R (1956d). Macroeconomics and linear programming. In *25 Economic Essays. In honour of Erik Lindahl 21 November 1956*, R Frisch (ed.), pp. 38–67. Stockholm: Ekonomisk Tidskrift, (Slightly simplified version of Frisch (1956a)).

Frisch, R (1957). The multiplex method for linear programming. *Sankhyá: The Indian Journal of Statistics* **18**, 329–362 [Practically identical to Frisch (1955)]

Gonzaga, CC (1992). Path-following methods for linear programming. *SIAM Review* **34**, 167–224.

Holt, CC, F Modigliani, JF Muth and HA Simon (1956). Derivation of a linear decision rule for production and employment. *Management Science* **2**, 159–177.

Hurwicz, L (1949). Linear programming and general theory of optimal behaviour (abstract). *Econometrica* **17**, 161–162.

Johansen, L (1952). Lineær programming. (Notes from Professor Ragnar Frisch's lecture 15 February 1952), pp. 15. Memorandum from Institute of Economics, University of Oslo, 17 February 1952, 15pp.

Karmarkar, NK (1984). A new polynomial-time algorithm for linear programming. *Combinatorica*, **4**, 373–395.

Klee, V and G Minty, (1972). How good is the simplex algorithm. In *Inequalities III*, O Sisha (ed.), pp. 159–175. New York, NY: Academic Press.

Koenker, R (2000). Galton, Edgeworth, Frisch, and prospects for quantile regression in econometrics. *Journal of Econometrics*, **95**(2), 347–374.

Kohli, MC (2001). Leontief and the bureau of labor statistics, 1941–54: developing a framework for measurement. In *The Age of Economic Measurement*, JL Klein and MS Morgan (eds.), pp. 190–212. Durham and London: Duke University Press.

Koopmans, T (1948). Optimum utilization of the transportation system (abstract). *Econometrica* **16**, 66.

Koopmans, TC (ed.) (1951). *Activity Analysis of Production and Allocation.* Cowles Commission Monographs No 13. New York: John Wiley & Sons.

Markowitz, H (1957). The elimination form of the inverse and its application in linear programming. *Management Science* **3**, 255–269.

Orden, A (1993). LP from the '40s to the '90s. *Interfaces* **23**, 2–12.

Sandmo, A (1993). Ragnar Frisch on the optimal diet. *History of Political Economy* **25**, 313–327.

Sharpe, W (1967). A linear programming algorithm for mutual fund portfolio selection. *Management Science* **13**, 499–511.

Stigler, G (1945). The cost of subsistence. *Journal of Farm Economics* **27**, 303–314.

Wood, MK and GB Dantzig (1949). Programming of interdependent activities: I general discussion. *Econometrica* **17**, 193–199.

Chapter 2

Methods of Modeling: Econometrics and Adaptive Control System

Topic 1

A Novel Method of Estimation Under Co-Integration

Alexis Lazaridis

Aristotle University of Thessaloniki, Greece

1. Introduction

We start with the unrestricted VAR(p):

$$\mathbf{x}_i = \boldsymbol{\delta} + \boldsymbol{\mu}\mathbf{t}_i + \sum_{j=1}^{p} \mathbf{A}_j \mathbf{x}_{i-j} + \mathbf{z}\mathbf{d}_i + \mathbf{w}_i \tag{1}$$

where $\mathbf{x} \in \mathrm{E}^n$ (here, E denotes the Euclidean space), t_i and d_i are the trend and dummy, respectively and $\mathbf{w}$ is the noise vector. The matrices of coefficients $\mathbf{A}_j$, are defined on $\mathrm{E}^n \times \mathrm{E}^n$. After estimation, we may compute matrix $\mathbf{\Pi}$ from:

$$\mathbf{\Pi} = \left(\sum_{j=1}^{p} \mathbf{A}_j \right) - \mathbf{I} \tag{1a}$$

or, directly, from the corresponding Error Correction VAR (ECVAR), which is presented later. Also, an estimate of matrix $\mathbf{\Pi}$ can be obtained by applying OLS, according to the procedure suggested by Kleibergen and Van Dijk (1994). It is re-called that this type of VAR models like the one presented above, have been recommended by Sims (1980), as a way to estimate the dynamic relationships among jointly endogenous variables.

It is known that matrix $\mathbf{\Pi}$ can be decomposed such that $\mathbf{\Pi} = \mathbf{AC}$, where the elements of $\mathbf{A}$ are the coefficients of adjustment and the rows of the matrix $\mathbf{C}$ are the (possible) co-integrating vectors. It should be clarified at the very outset, that in many books etc., $\mathbf{C}$ is denoted by $\boldsymbol{\beta}'$ (prime shows transposition) and $\mathbf{A}$ by $\boldsymbol{\alpha}$, although capital letters are used for denoting matrices. Besides, in the general linear model, $\mathbf{y} = \mathbf{X}\boldsymbol{\beta} + \mathbf{u}$, $\boldsymbol{\beta}$ is the vector of coefficients. To avoid any confusion, we adopted the notation used here.

A straightforward approach to obtain matrices $\mathbf{A}$ and $\mathbf{C}$, provided that some rank conditions of $\mathbf{\Pi}$ are satisfied, is to solve the eigenproblem (Johnston and Di Nardo, 1997, pp. 287–296).

$$\mathbf{\Pi} = \mathbf{V\Lambda V}^{-1} \tag{2}$$

where $\mathbf{\Lambda}$ is the diagonal matrix of eigenvalues and $\mathbf{V}$ is the matrix of the corresponding eigenvectors. Note that since $\mathbf{\Pi}$ is not symmetric, $\mathbf{V} \neq \mathbf{V}^{-1}$. Eq. (2) holds if $\mathbf{\Pi}$ is $(n \times n)$ and all its eigenvalues are real, which implies that the eigenvectors are real too. In cases that $\mathbf{\Pi}$ is defined on $E^n \times E^m$, with $m > n$, Eq. (2) is not applicable. If these necessary conditions hold, then from Eq. (2) follows that matrix $\mathbf{A} = \mathbf{V\Lambda}$ and $\mathbf{C} = \mathbf{V}^{-1}$. Further, the co-integrating vectors are normalized so that — usually — the first element to be equal to 1. In other words, if $c_1^i\,(\neq 0)$, denotes the first element[1] of the ith row of matrix $\mathbf{C}$, i.e.,[2] $\mathbf{c}'_{i.}$ and $\mathbf{a}_j$ the jth column of matrix $\mathbf{A}$, then the normalization process can be summarized as $\mathbf{c}'_{i.}/\mathbf{c}_1^i$ and $\mathbf{a}_i \times \mathbf{c}_1^i$, for $i = 1, \ldots, n$.

It is known that according to the maximum likelihood (ML) method, matrix $\mathbf{C}$ can be obtained by solving the constrained problem:

$$\{\min \det(\mathbf{I} - \mathbf{C}\hat{\mathbf{\Sigma}}_{k0}\hat{\mathbf{\Sigma}}_{00}^{-1}\hat{\mathbf{\Sigma}}_{0k}\mathbf{C}') \mid \mathbf{C}\hat{\mathbf{\Sigma}}_{kk}\mathbf{C}' = \mathbf{I}\} \tag{3}$$

where matrices $\hat{\mathbf{\Sigma}}_{00}$, $\hat{\mathbf{\Sigma}}_{0k}$, $\hat{\mathbf{\Sigma}}_{kk}$ and $\hat{\mathbf{\Sigma}}_{k0}$ are residual co-variance matrices of specific regressions (Harris, 1995, p. 78; Johansen Juselius, 1990). These matrices appear in the last term of the concentrated likelihood of[3] Eq. (5) together with $\mathbf{C}$, where $\mathbf{\Pi}$ is replaced by $\mathbf{AC}$.

Note that the minimization of Eq. (3) is with respect to the elements of matrix $\mathbf{C}$. It is re-called that Eq. (3) is minimized by solving a general eigenvalue problem, i.e., finding the eigenvalues from:

$$|\boldsymbol{\lambda}\hat{\mathbf{\Sigma}}_{kk} - \hat{\mathbf{\Sigma}}_{k0}\hat{\mathbf{\Sigma}}_{00}^{-1}\hat{\mathbf{\Sigma}}_{0k}| = 0. \tag{4}$$

Applying Cholesky's factorization on the positive definite symmetric matrix $\hat{\mathbf{\Sigma}}_{kk}^{-1}$, such that $\hat{\mathbf{\Sigma}}_{kk}^{-1} = \mathbf{LL}' \Rightarrow \hat{\mathbf{\Sigma}}_{kk} = (\mathbf{L}')^{-1}\mathbf{L}^{-1}$, where $\mathbf{L}$ is lower

---

[1] In some computer programs, $\mathbf{c}_i^i$ is taken to be the ith element of the ith row.

[2] Note that dot is necessary to distinguish the jth row of $\mathbf{C}$, i.e., $\mathbf{c}'_{i.}$, from the transposed of the jth column of this matrix (i.e., $\mathbf{c}'_i$).

[3] Equation (5) is presented below.

triangular, Eq. (4) takes the conventional form[4]:

$$|\lambda \mathbf{I} - \mathbf{L}'\hat{\mathbf{\Sigma}}_{k0}\hat{\mathbf{\Sigma}}_{00}^{-1}\hat{\mathbf{\Sigma}}_{0k}\mathbf{L}| = 0. \tag{4a}$$

After obtaining matrix $\mathbf{V} = \mathbf{V}'$ of the eigenvectors, we can easily compute the (unormalized) $\mathbf{C}$ from $\mathbf{C} = \mathbf{V}'\mathbf{L}'$. Note also that $\mathbf{C}\hat{\mathbf{\Sigma}}_{kk}\mathbf{C}' = \mathbf{V}'\mathbf{L}'(\mathbf{L}')^{-1}\mathbf{L}^{-1}\mathbf{L}\mathbf{V} = \mathbf{I}$. Finally, matrix $\mathbf{A}$ is obtained from $\mathrm{A} = \hat{\mathbf{\Sigma}}_{0k}\mathbf{C}'$. Hence, the initial problem reduces to finding one of the eigen values and eigenvectors of a symmetric matrix, which are real. And this is the main advantage of the ML method. According to this procedure, a constant and/or trend can be included in the co-integrating vectors, but it is not clear of how to accommodate additional deterministic factors, such as one or more dummies. Besides, nothing is said about the variance of the (disequilibrium) errors, which correspond to the finally adopted co-integration vector.

We propose in the first part of this chapter, a comparatively simple method, to directly obtain co-integrating vectors, which may include, apart from the standard deterministic components, any number of dummies, in cases that this is necessary. Additionally, as we will see in what follows, the errors corresponding to the proper co-integration vector have the property of minimum variance.

2. Some Estimation Remarks

It was mentioned that after estimation, matrix $\mathbf{\Pi}$ can be computed from Eq. (1a). It is obvious that each equation of the VAR can be estimated by OLS. At this stage, one should be aware to comply with the basic assumptions regarding the error terms, in the sense that proper tests must ensure that the noises are white. This way, the order p of the model is defined imposing at the same time restrictions — if necessary — on some elements of $\mathbf{A}$'s and vectors $\boldsymbol{\mu}$ and $\mathbf{z}$ to be zero.

It can be easily shown that Eq. (1) may be expressed in the form of an equivalent ECVAR, i.e.,

$$\Delta\mathbf{x}_i = \boldsymbol{\delta} + \boldsymbol{\mu}t_i + \sum_{j=1}^{p-1}\mathbf{Q}_j\Delta\mathbf{x}_{i-j} + \mathbf{\Pi}\mathbf{x}_{i-p} + \mathbf{z}d_i + \mathbf{w}_i \tag{5}$$

[4]It should be noted that it is also necessary to pre-multiply the elements of Eq. (4) by $\mathbf{L}'$ and post-multiply them by $\mathbf{L}$.

where

$$\mathbf{Q}_j = \left(\sum_{k=1}^{j} \mathbf{A}_k - \mathbf{I} \right).$$

3. The Singular Value Decomposition

Matrix $\mathbf{\Pi}$ in Eq. (5) may be augmented to accommodate, as an additional column, the vector $\boldsymbol{\delta}$, $\boldsymbol{\mu}$ or $\mathbf{z}$. It is also possible to accommodate all these vectors in the following way:

$$\tilde{\mathbf{\Pi}} = \left[\mathbf{\Pi} \vdots \boldsymbol{\delta\mu}\mathbf{z}\right] \tag{6}$$

In this case, $\tilde{\mathbf{\Pi}}$ is defined on $E^n \times E^m$, with $m > n$.

For conformability, the vector $\mathbf{x}_{i-p}$ in Eq. (5) should be augmented accordingly by using the linear advance operator L (such that $\mathrm{L}^k y_i = y_{i+k}$), i.e.,

$$\tilde{\mathbf{x}}_{i-p} = \begin{bmatrix} \mathbf{x}_{i-p} \\ 1 \\ L^p t_{i-p} \\ L^p d_{i-p} \end{bmatrix}.$$

Hence, Eq. (5) takes the form:

$$\Delta \mathbf{x}_i = \sum_{j=1}^{p-1} \mathbf{Q}_j \Delta \mathbf{x}_{i-j} + \tilde{\mathbf{\Pi}} \tilde{\mathbf{X}}_{i-p} + \mathbf{w}_i. \tag{7}$$

We can compute the unique generalized inverse (or pseudo-inverse) of $\tilde{\mathbf{\Pi}}$, denoted by $\tilde{\mathbf{\Pi}}^+$ (Lazaridis and Basu, 1981), from:

$$\tilde{\mathbf{\Pi}}^+ = \mathbf{VF}^*\mathbf{U}' \tag{8}$$

where

$\mathbf{V}$ is of dimension $(m \times m)$, and its columns are the orthonormal eigenvectors of $\tilde{\mathbf{\Pi}}'\tilde{\mathbf{\Pi}}$.

$\mathbf{U}$ is $(n \times m)$, and its columns are the orthonormal eigenvectors of $\tilde{\mathbf{\Pi}}\tilde{\mathbf{\Pi}}'$, corresponding to the largest m eigenvalues of this matrix so that:

$$\mathbf{U}'\mathbf{U} = \mathbf{V}'\mathbf{V} = \mathbf{VV}' = \mathbf{I}_m \tag{9}$$

$\mathbf{F}$ is diagonal $(m \times m)$, with elements $f_{ii} \triangleq f_i$, called the singular values of $\tilde{\mathbf{\Pi}}$, which are the non-negative square roots of the eigenvalues of $\tilde{\mathbf{\Pi}}'\tilde{\mathbf{\Pi}}$.

If the rank of $\tilde{\mathbf{\Pi}}$ is k, i.e.,

$$r(\tilde{\mathbf{\Pi}}) = k \leq n, \quad \text{then } f_1 \geq f_2 \geq \cdots \geq f_k > 0$$

and $f_{k+1} = f_{k+2} = \cdots = f_m = 0$

$\mathbf{F}^*$ is diagonal $(m \times m)$, and $f^*_{ii} \triangleq f^*_i = 1/f_i$.

It should be noted that all the above matrices are real. Also note that if $\tilde{\mathbf{\Pi}}$ has full row rank, then $\tilde{\mathbf{\Pi}}^+$ is the right inverse of $\tilde{\mathbf{\Pi}}$, i.e., $\tilde{\mathbf{\Pi}}\tilde{\mathbf{\Pi}}^+ = I_n$

Hence, the singular value decomposition (SVD) of $\tilde{\mathbf{\Pi}}$ is:

$$\tilde{\mathbf{\Pi}} = \mathbf{UFV}'. \tag{10}$$

4. Computing the Co-Integrating Vectors

We proceed to form a matrix $\mathbf{F}_1$ of dimension $(m \times n)$ such that:

$$f_{1(i,j)} = \begin{cases} 0, & \text{if } i \neq j \\ \sqrt{f_i}, & \text{if } i = j \end{cases} \tag{11}$$

In accordance to Eq. (11), we next form a matrix $\mathbf{F}_2$ of dimension $(n \times m)$. It is verified that $\mathbf{F}_1\mathbf{F}_2 = \mathbf{F}$. Hence, Eq. (10) can be written as:

$$\tilde{\mathbf{\Pi}} = \mathbf{AC}$$

where $\mathbf{A} = \mathbf{UF}_1$ of dimension $(n \times n)$ and $\mathbf{C} = \mathbf{F}_2\mathbf{V}$ of dimension $(n \times m)$.

After normalization, we get the possible co-integrating vectors as the rows of matrix $\mathbf{C}$, whereas the elements of $\mathbf{A}$ can be viewed as approximations to the corresponding coefficients of adjustment.

It is apparent that the co-integrating vectors can always be computed, given that $r(\tilde{\mathbf{\Pi}}) > 0$. Needless to say that the same steps are followed if $\mathbf{\Pi}$, instead of $\tilde{\mathbf{\Pi}}$, is considered.

5. Case Study 1: Comparing Co-Integrating Vectors

We first consider the VAR(2) model, presented by Holden and Peaman (1994). The $\mathbf{x} \sim I(1)$ vector consists of the variables C (consumption), I (income) and W (wealth), so that $\mathbf{x} \in E^3$. We used the same set of data presented at the end of this book (Table D3, pp. 196–198). To simplify the presentation, we did not consider the three dummies DD682, DD792, and DD883. Hence, the VAR in this case has a form analogous to Eq. (1), without trend and dummy components.

Estimating the equations of this VAR by OLS, we get the following results.

$$\mathbf{\Pi} = \begin{bmatrix} 0.059472 & -0.073043 & 0.0072427 \\ 0.5486237 & -0.5244858 & -0.028041 \\ 0.3595042 & -0.2938160 & -0.039400 \end{bmatrix}, \quad \boldsymbol{\delta} = \begin{bmatrix} 0.064873 \\ 0.16781451 \\ -0.1554421 \end{bmatrix}. \tag{12}$$

Hat (^) is omitted for simplicity. One possible co-integration vector with intercept, obtained by applying the ML method is:

$$1 - 0.9574698 - 0.048531 + 0.2912925 \tag{13}$$

and corresponds to the long-run relationship:

$$C_i = 0.9574698 I_i + 0.048531 W_i - 0.2912925.$$

For the vector in Eq. (13) to be a co-integration vector, the necessary and sufficient condition is that the (disequilibrium) errors $\hat{u}_i$ computed from:

$$\hat{u}_i = C_i - 0.9574698 I_i - 0.048531 W_i + 0.2912925 \tag{14}$$

to be stationary, i.e., $\{\hat{u}_i\} \sim I(0)$, since $\mathbf{x} \sim I(1)$.

For distinct purposes, these ML errors obtained from Eq. (14) will be denoted by uml_i.

Applying SVD on $\tilde{\mathbf{\Pi}} = [\mathbf{\Pi} \vdots \delta]$ from Eq. (12), we get matrix $\mathbf{C}$ which is:

$$\mathbf{C} = \begin{bmatrix} 1 & -0.9206225 & -0.06545528 & 0.1110597 \\ 1 & 0.3586208 & -0.6305243 & -6.403011 \\ 1 & 1.283901 & -2.050713 & 0.4300255 \end{bmatrix}$$

$$\text{Euclidean norm} = \begin{pmatrix} 1.365344 \\ 6.521098 \\ 2.653064 \end{pmatrix}.$$

In the column at the right-hand side, the (Euclidean) norm of each row of $\mathbf{C}$ is presented. It is noted also that the singular values of $\mathbf{\Pi}$ are:

$$f_1 = 0.8979247, \quad f_2 = 0.2304381 \quad \text{and} \quad f_3 = 0.0083471695.$$

All f_i are less than 1, indicating thus that at least one row of $\mathbf{C}$ can be regarded as a possible co-integration vector, in the sense that the resulting errors are likely to be stationary. This row usually has the smallest norm, which in this case, is the first one. Hence, the errors corresponding to this co-integration vector are obtained from:

$$\hat{u}_i = C_i - 0.9206225 I_i - 0.06545528 W_i + 0.1110597. \qquad (15)$$

These errors will be denoted by $usvd_i$. It should be pointed out here that the smallest norm condition is just an indication as to which row of matrix **C** to be considered first.

It may be useful to re-call at this point that the co-integration vector reported by the authors (p. 111), which is obtained by applying OLS after omitting the first two observations from the initial sample is:

$$1 - 0.9109 - 0.0790 + 0.1778.$$

It should be clarified that this vector has been obtained by applying OLS to estimate directly the long-run relationship. Repeating this procedure, we get:

$$1 - 0.910971 - 0.079761 + 0.178455.$$

Hence, the errors are computed from:

$$\hat{u}_i = C_i - 0.910971 I_i - 0.079761 W_i + 0.178455. \qquad (16)$$

These errors will be denoted by $ubook_i$.

The next task is to compare the above three series of errors, having a first look at the corresponding graphs, which are shown in Fig. 1.

Although by a first glance the three series seem to be stationary, we shall go a bit further to see their properties in detail. Before going to the next section, it is worth to mention here that for the three series, the normality test using the Jarque-Bera criterion favors the null.

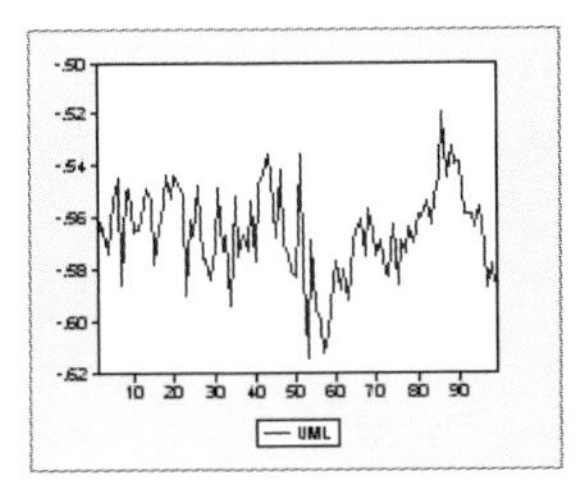

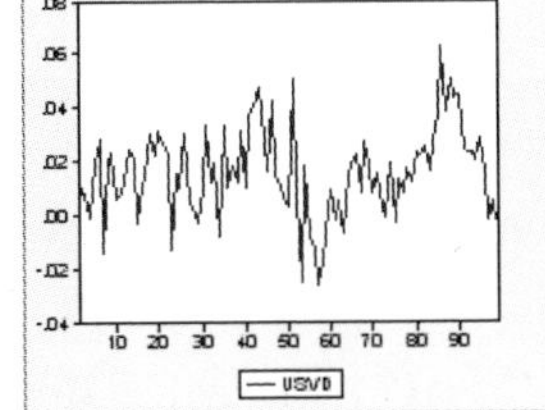

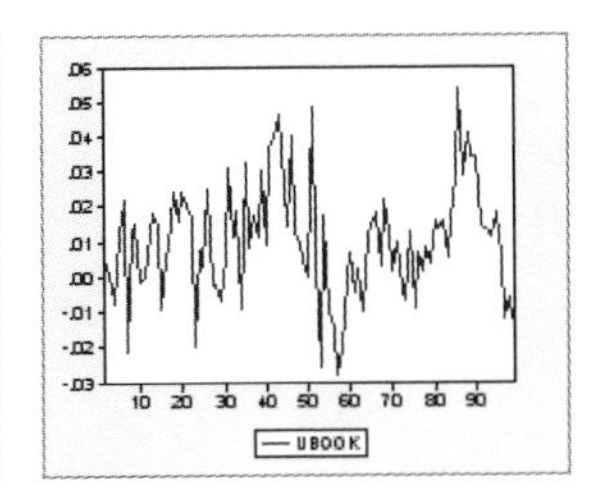

Figure 1. The series $\{uml_i\}$, $\{usvd_i\}$, and $\{ubook_i\}$.

5.1. *Testing for Co-integration*

Since for the first two cases, $\sum \hat{u}_i \neq 0$, we estimated the following regression:

$$\Delta\hat{u}_i = b_1 + b_2\hat{u}_{i-1} + \sum_{j=1}^{q} b_{j+2}\Delta\hat{u}_{i-j} + \varepsilon_i. \tag{17}$$

The value of q was set such that the noises ε_i to be white, as it will be explained in detail later on. Regarding the last case, the intercept is omitted from Eq. (17), since $\hat{u}_i$'s are OLS residuals, so that $\sum \hat{u}_i = 0$.

The estimation results are as follows:

$$\Delta u\hat{m}l_i = -0.00612 - \underset{(0.193688)}{0.381773} uml_{i-1} - 0.281651\Delta uml_{i-1}$$

$$t = -3.682$$

$$\Delta u\hat{s}vd_i = 0.006525 - \underset{(0.108269)}{0.428278} usvd_{i-1} - 0.259066\Delta usvd_{i-1}$$

$$t = -3.956$$

$$\Delta ub\hat{o}ok_i = -\underset{(0.09522)}{0.606759} ubook_{i-1}$$

$$t = -6.37.$$

Since these error series are the results of specific calculations and in the simplest case are the OLS residuals, it is not advisable (Harris, 1995, pp. 54–55). to apply the Dickey-Fuller (DF/ADF) test, as we do with any variable in the initial data set. We have to compute the t_u statistic from McKinnon (1991, p. 267–276) critical values (see also Harris, 1995, Table A6, p. 158). This statistic ($t_u = \Phi_\infty + \Phi_1/\mathrm{T} + \Phi_2/\mathrm{T}^2$) for three (independent) variables in the long-run relationship with intercept is:

$\alpha = 0.01$	$\alpha = 0.05$	$\alpha = 0.10$
$t_u = -4.45$	$t_u = -3.83$	$t_u = -3.52$.

Hence, {book$_i$} is stationary for $\alpha = (0.01, 0.05, 0.10)$, $\{usvd_i\}$ is stationary for α = (0.05, 0.10) and $\{uml_i\}$ is stationary only for $\alpha = 0.10$. Recall that the null $[\hat{u}_i \sim \mathrm{I}(1)]$ is rejected in favor of $H_1[\hat{u}_i \sim I(0)]$, if $t < t_u$. Besides, if we compute the standard deviation of these three series we will get:

Standard deviation of $\{uml_i\}$	Standard deviation of $\{usvd_i\}$	Standard deviation of $\{book_i\}$
0.017622	0.016612	0.0164

It is obvious from these findings, the super consistency of the OLS. However, the method we introduce here, produces almost minimum variance errors, which is a quite desirable property, as suggested by Engle and Granger (Dickey *et al.*, 1994, p. 27). Hence, we will mainly focus on this property in our next section.

6. Case Study 2: Further Evidence Regarding the Minimum Variance Property

With the same set of data, the co-integration vector without an intercept obtained by applying the ML method is:

$$1 - 0.9422994 - 0.058564$$

The errors corresponding to this vector are obtained from:

$$\hat{u}_i = C_i - 0.9422994 I_i - 0.058564 W_i \tag{18}$$

Again these errors will be denoted by uml_i.

Next, we apply SVD to $\mathbf{\Pi}$ in Eq. (12) to obtain matrix $\mathbf{C}$, which is:

$$\mathbf{C} = \begin{bmatrix} 1 & -0.9192042 & -0.066081211 \\ 1 & 1.160720 & -1.012970 \\ 1 & 0.9395341 & 2.063768 \end{bmatrix}$$

and

$$\text{Euclidean norm} = \begin{pmatrix} 1.359891 \\ 1.836676 \\ 2.47828 \end{pmatrix}$$

The singular values of $\mathbf{\Pi}$ are:

$$f_1 = 0.89514, \; f_2 = 0.0403121 \; \text{ and } \; f_3 = 0.0026218249$$

Again, all f_i's are less than 1. We consider the first row of $\mathbf{C}$, so that the errors corresponding to this vector are obtained from:

$$\hat{u}_i = C_i - 0.9192042 I_i - 0.066081211 W_i \tag{19}$$

These errors will be also denoted by $usvd_i$.

The co-integrating vector reported by the authors (p. 109), which is obtained from the VAR(2), including the dummies mentioned previously, by applying the ML procedure is given by:

$$1 - 0.93638 - 0.03804$$

and the errors obtained from this vector are computed from:

$$\hat{u}_i = C_i - 0.93638 I_i - 0.03804 W_i \tag{20}$$

As in the previous case, these errors will be denoted by $ubook_i$.

The minimum variance property can be seen from the following:

Standard deviation of $\{uml_i\}$	Standard deviation of $\{usvd_i\}$	Standard deviation of $\{book_i\}$
0.0169	0.0166	0.0193

Next, we consider the model presented by Banerjee *et al.* (1993, pp. 292–293) that refers to VAR(2) with constant dummy and trend. The state vector $\mathbf{x} \in E^4$ consists of the variables $(m - p)$, Δp, y, and R. The data (in logs) are presented in Harris (1995, statistical appendix, pp. 153–155. Note that the entries for 1967:1 and 1973:2 have to be corrected from 0.076195 and 0.56569498 to 9.076195 and 9.56569498, respectively). Considering the co-integration vector reported by the authors, which has been obtained by applying the ML method, we compute the errors from:

$$\hat{u}_i = (m - p)_i + 6.3966 \Delta p_i - 0.8938 y_i + 7.6838 R_i. \tag{21}$$

These errors will be denoted by uml_i.

Estimating the same VAR by OLS, we get,

$$\mathbf{\Pi} = \begin{bmatrix} -0.089584 & 0.16375 & -0.184282 & -0.933878 \\ -0.012116 & -0.9605 & 0.229635 & 0.206963 \\ 0.021703 & 0.676612 & -0.476152 & 0.447157 \\ -0.006877 & 0.118891 & 0.048932 & -0.076414 \end{bmatrix},$$

$$\boldsymbol{\delta} = \begin{bmatrix} 3.129631 \\ -2.46328 \\ 5.181506 \\ -0.470803 \end{bmatrix} \quad \text{and} \quad \boldsymbol{\mu} = \begin{bmatrix} 0.001867 \\ -0.001442 \\ 0.003078 \\ -0.000366 \end{bmatrix}$$

Following the procedure described above, we obtain matrix **C** which is:

$$\mathbf{C} = \begin{bmatrix} 1 & -48.30517 & 24.60144 & 37.75685 \\ 1 & 5.491415 & -0.6510755 & 7.423318 \\ 1 & -2.055229 & -5.466846 & 0.9061699 \\ 1 & 0.0051181894 & 0.1603713 & -0.1244312 \end{bmatrix}$$

$$\text{Euclidean norm} = \begin{pmatrix} 66.06966 \\ 9.310488 \\ 5.99429 \\ 1.020406 \end{pmatrix}$$

The singular values of $\mathbf{\Pi}$ are:

$$f_1 = 1.503069, \; f_2 = 0.7752298, \; f_3 = 0.1977592 \\ \text{and} \quad f_4 = 0.012560162$$

The fact that one singular value is greater than one, is an indication that unless a dummy is present — we hardly can trace one row of matrix **C**, which may be related to stationary errors $\hat{u}_i$. As shown in Fig. 1, from the graphs, from which refer to the errors corresponding to the rows of **C**, we see that only the second row produces reasonably results. In this case, the errors are computed from:

$$\hat{u}_i = (m - p)_i + 5.491415\Delta p_i - 0.6510755 y_i + 7.42318 R_i. \qquad (22)$$

As usual, these errors will be denoted by $usvd_i$.
The minimum variance property can be seen from:

Standard deviation of $\{uml_i\}$	Standard deviation of $\{usvd_i\}$
0.2671	0.2375

To see that none of the two error series is stationary, as it was indicated by the first singular value, we present the following results.

$$\Delta u\hat{m}l_i = 0.268599 - \underset{(0.065634)}{0.178771}\, uml_{i-1} - 0.214098\Delta uml_{i-1}$$
$$t = -2.724$$
$$\Delta u\hat{s}vd_i = 0.832329 - \underset{(0.066421)}{0.193432}\, usvd_{i-1} - 0.160135\Delta usvd_{i-1}$$
$$t = -2.912$$

For a long-run relationship with three (independent) variables without trend and intercept, the critical value t_u for $\alpha = 0.10$ is less than -3.

Nevertheless, the method proposed, produces better results than the ML approach does.

Finally, we will consider the first of the two co-integration vectors (since, it produces slightly better results), reported by Harris (1995, p. 102). From this vector, which was obtained by the ML method, one gets the following errors.

$$\hat{u}_i = (m - p)_i + 7.325\Delta p_i - 1.073y_i + 6.808R_i \tag{23}$$

Also we obtain

standard deviation of $\hat{u}_i$'s $= 0.268$ and

$$\Delta\hat{\hat{u}}_i = -0.117378 - \underset{(0.068296)}{0.178896}\hat{u}_{i-1} - 0.30095417\Delta\hat{u}_{i-1}$$

$$t = -2.619.$$

It is obvious that these additional results further support the previous findings, which are due to some properties of SVD (Lazaridis, 1986).

6.1. *A Note on Integration*

It seems to be a common belief that differentiating a variable say n times, we will always get a stationary series (Harris, 1995, p. 16). Hence, the initial series is said to be $I(n)$. But, this is not necessarily the case, as it is verified that if we take into consideration the exports of goods and services for Greece, as presented in Table 1. Besides if n is large enough, is it of any good to obtain such a stationary series when economic variables are considered?

Table 1. The series $\{x_i\}$.

Exports of goods and services in bil. Eur. Greece 1956–2000.				
X				
2.4944974E-02	2.9053558E-02	2.9347029E-02	2.8760089E-02	2.8173149E-02
3.2575201E-02	3.5803374E-02	4.1379310E-02	4.2553190E-02	4.7248717E-02
6.6030815E-02	6.7498162E-02	6.6030815E-02	7.6008804E-02	8.8041089E-02
0.1000734	0.1300073	0.2022010	0.2664710	0.3325018
0.4258254	0.4763023	0.5998532	0.7325019	1.049743
1.239618	1.388114	1.788701	2.419956	2.868965
3.618782	4.510052	4.978430	5.818929	6.485106
7.690976	9.316215	9.847396	11.45708	14.08716
15.39428	18.87601	20.90506	22.56992	25.51577

Source: International Financial Statistics. (International Monetary Fund, Washington DC)

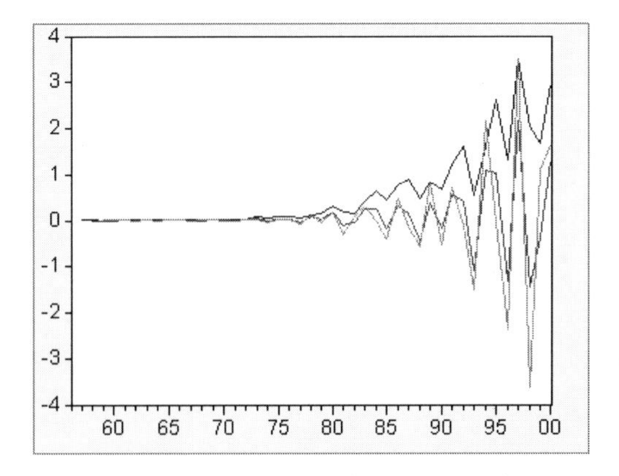

The series $\{\Delta x_i\}$, $\{\Delta^2 x_i\}$, and $\{\Delta^3 x_i\}$

Figure 2. Differences of the NI series $\{x_i\}$.

Using the series $\{x_i\}$ as shown in Table 1, we can easily verify from the corresponding graph, that even for $n = 5$, the series $\{\Delta^5 x_i\}$ is not stationary. In Fig. 2, the series $\{\Delta x_i\}$, $\{\Delta^2 x_i\}$, and $\{\Delta^3 x_i\}$ are presented, for a better understanding of this situation. We will call the variables, which belong to this category near the integrated (NI) (Banerjee *et al.*, 1993, p. 95) series.

To trace that a given series is NI, we may run the regression:

$$\Delta x_i = \beta_1 + \beta_2 x_{i-1} + \beta_3 t_i + \sum_{j=1}^{q} \beta_{j+3} \Delta x_{i-j} + u_i \tag{24}$$

which is a re-parametrized AR(q) with constant, where we added a trend too. As shown in Eq. (17), the value of q is set such that the noises u_i to be white. A comparatively simple way for this verification is to compute the residuals $\hat{u}_i$ and to consider the corresponding Ljung-Box Q statistics and particularly their p-values, which should be much greater than 0.1, to say that no autocorrelation (AC) is present. On the other hand, if we trace heteroscedasticity, this is a strong indication that we have a NI series. For the data presented in Table 1, we found that $q = 2$. The corresponding Q statistics (Column 4) together with p-values are presented in Table 2.

We see that for all k (column 1), the corresponding p-values (column 5) are greater than 0.1. But, if we look at the residuals graph (Fig. 3), we can verify the presence of heteroscedasticity.

Table 2. Residuals: Autocorrelations, PAC, Q statistics and *p*-values.

Autocorrelation (AC), partial autocor coefficients (PAC) Q statistics and p values (prob).						
Values of k	AC	PAC	Q_Stat(L-B)	p value	Q_Stat(B-P)	p value
1	0.040101	0.040101	0.072482	0.787756	0.067540	0.794952
2	−0.061329	−0.063039	0.246254	0.884151	0.225515	0.893367
3	−0.008189	−0.003064	0.249432	0.969240	0.228331	0.972891
4	−0.285494	−0.290504	4.213238	0.377916	3.651618	0.455202
5	0.042856	0.072784	4.304969	0.506394	3.728756	0.589090
6	0.099724	0.058378	4.815469	0.567689	4.146438	0.656867
7	−0.151904	−0.167680	6.033816	0.535806	5.115577	0.645861
8	−0.007846	−0.069642	6.037162	0.643069	5.118163	0.744875
9	0.001750	0.023203	6.037333	0.736176	5.118291	0.823877
10	0.109773	0.165620	6.733228	0.750367	5.624397	0.845771
11	0.134538	0.022368	7.812250	0.730017	6.384616	0.846509
12	−0.029101	−0.048363	7.864417	0.795634	6.420185	0.893438
13	−0.074997	−0.028063	8.222832	0.828786	6.656414	0.918979
14	−0.084284	−0.017969	8.691681	0.850280	6.954772	0.936441
15	−0.113984	−0.104484	9.580937	0.845240	7.500452	0.942248
16	−0.081662	−0.157803	10.05493	0.863744	7.780540	0.955133
17	−0.001015	−0.005026	10.05501	0.901288	7.780582	0.971016
18	−0.053178	−0.056574	10.27276	0.922639	7.899356	0.980097

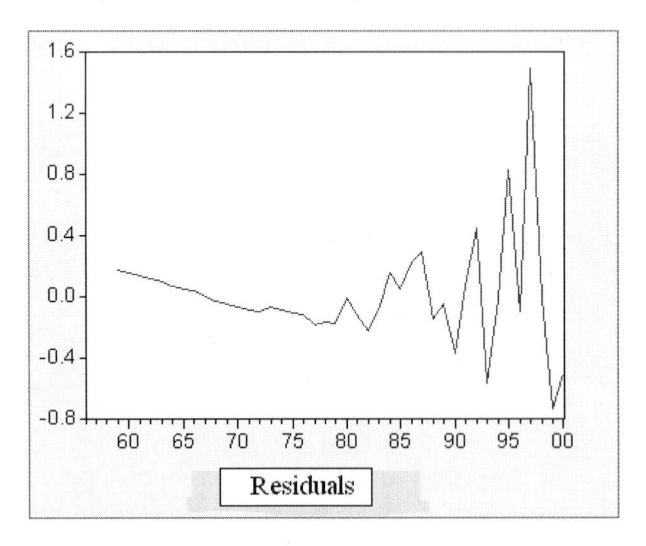

Figure 3. The residuals from Eq. (24) ($q = 2$).

Another easy and practical way to trace heteroscedasticity, is to select the explanatory variable, which yields the smallest p-value for the corresponding Spearman's correlation coefficient (r_s). Note that in models like the one we have seen in Eq. (24), such a variable is the trend, resulting in this case to: $r_s = 0.3739, t = 2.394, p = 0.0219$. This means that for a significance level $\alpha > 0.022$, heteroscedasticity is present. Hence, we have an initially NI series $\{x_i\}$, which has to be weighed somehow in order to be transformed to a difference stationary series (DSS). In some cases, this transformation can be achieved if the initial series is expressed in real terms, or as percentage of growth. In usual applications, the (natural) logs [i.e., $ln(x_i)$] are considered, given that $\{\Delta x_i/x_i\} > \{\Delta ln(x_i)\} > \{\Delta x_i/x_{i-1}\}$.

7. Case Study 3: How to Obtain Reliable Results Regarding the Order of Integration?

It is a common practice to apply DF/ADF for determining the order of integration for a given series. However, the analytical step-by-step application of this test (Enders, 1995, p. 257; Holden and Perman, 1994, pp. 64–65), leaves the impression that the cure is worse than the disease itself. On the other hand, by simply adopting the results produced by some commercial computer programs, which is the usual practice in many applications, we may reach some faulty conclusions. It should be noted at this point that according to the results obtained by a well-known commercial package, the series $\{x_i\}$, as shown in Table 1, should be considered as I(2), for $\alpha = 0.05$, which is not the case.

Here, we present a comparatively simple procedure to determine the order of integration of a DSS, by inspecting a model analogous to Eq. (24). Assuming that $\{z_i\}$ is DSS, we proceed as follows:

- Start with $k = 1$ and estimate the regression:

$$\Delta^k z_i = \beta_1 + \beta_2 \Delta^{k-1} z_{i-1} + \beta_3 t_i + \sum_{j=1}^{q} \beta_{j+3} \Delta^k z_{i-j} + u_i. \tag{25}$$

Be sure that the noises in Eq. (25) are white, by applying the relevant tests as described above. This way, we set the lag-length q. Note that if $k = 1$, its value is omitted from Eq. (25) and that $\Delta^0 z_{i-1} = z_{i-1}$.

Table 3. Critical values for the DF F-test.

Sample size	$\alpha = 0.01$	$\alpha = 0.05$	$\alpha = 0.10$
25	10.61	7.24	5.91
50	9.31	6.73	5.61
100	8.73	6.49	5.47
250	8.43	6.34	5.39
500	8.34	6.30	5.36
>500	8.27	6.25	5.34

- Proceed to test the hypothesis:

$$H_0 : \beta_2 = \beta_3 = 0. \tag{26}$$

To compare the computed F-statistic, we have to consider, in this case, the Dickey-Fuller F-distribution. The critical values (Dickey and Fuller, 1981) are reproduced in Table 3, to facilitate the presentation.

Note that in order to reject Eq. (26), F-statistic should be much greater than the corresponding critical values, as shown in Table 3. If we reject Eq. (26), this means that the series $\{z_i\}$ is I$(k-1)$. If the null hypothesis is not rejected, then increase k by $1(k = k+1)$ and repeat the same procedure, i.e., starting from estimating Eq. (25), testing at each stage so that the model disturbances are white noises.

- In case of known structural break(s) in the series, a dummy d_i should be added in Eq. (25), such that:

$$d_i = \begin{cases} 0 & \text{if } i \leq r \\ w_i & \text{if } i > r \end{cases}$$

where $w_i = 1(\forall i)$ for the shift in mean, or $w_i = t_i$ for the shift in trend. It is re-called that r denotes the date of the break or shift. With this specification, the results obtained are alike the ones that Perron *et al.* (1992a; b) procedure yields.

We applied the proposed technique to find the order of integration of the variable m_i as shown in Eq. (21), which is the log of nominal M1 (money supply). Harris (1995, p. 38) after applying DF/ADF test reports that "... *and the interest rate as* $I(1)$ *and prices as* (I2). *The nominal money supply might be either, given its path over time....*". Since the latter variable is m_i, it is clear that, with this particular test, it was not possible to get a precise answer as to whether $\{m_i\}$ is either $I(1)$ or $I(2)$.

To show the details, underline the first stage of this procedure; we present some estimation results with different values of q.

For $q = 3$

Most of the p-values in the 5th column of the table, which is analogous to Table 2, are equal to zero.

Obviously, this is not the proper lag-length.

For $q = 4$

All p-values are greater than 0.1, as shown in Table 2. $r_s = -0.193$, $p = 0.05652$. $F(2, 94) = 4.2524$.

This lag-length is acceptable.

For $q = 5$

All p-values are greater than 0.1, as shown in Table 2. $r_s = -0.2094$, $p = 0.0395.F(2, 92) = 4.2278$.

This lag-length is questionable, since for $\alpha > 0.04$, we face the problem of heteroscedasticity.

$q = 6$

All p-values are greater than 0.1, as we have seen in Table 2. $r_s = -0.166$, $p = 0.1038.F(2, 90) = 5.937$.

This lag-length is quite acceptable.

$q = 7$

All p-values are greater than 0.1, as shown in Table 2. $r_s = -0.1928$, $p = 0.0608.F(2, 88) = 4.768$.

This lag-length is also acceptable, but it is inferior when compared to the previous one.

From these analytical results, it can be easily verified that the proper lag-length is 6 ($q = 6$). It may also be worthy to mention that according to the normality test (Jarque-Bera), we should accept the null. The value of $F = 5.937$ is less than the corresponding critical values as shown in Table 3, for 100 observations and $\alpha = (0.01, 0.05)$. Hence, we accept Eq. (26) and increase the value of k by one. The estimation results, after properly selecting the value of $q(q = 5)$, are:

$$\Delta^2 m_i = \underset{(0.0046)}{0.0047} - \underset{(0.21784)}{0.894892}\,\Delta m_{i-1} + \underset{(0.000108)}{0.00033}\,t_i + \sum_{j=1}^{5} \hat{\beta}_{j+3}\Delta^2 m_{i-j} + \hat{u}_i. \tag{27}$$

Regarding the residuals, all p-values in the 5th column of the table, which is analogous to Table 2, are greater than 0.1. Further, we have: $r_s = -0.1556$, $p = 0.127.$ Jarque-Bera$= 0.8(p = 0.67).F(2, 91) = 8.44$.

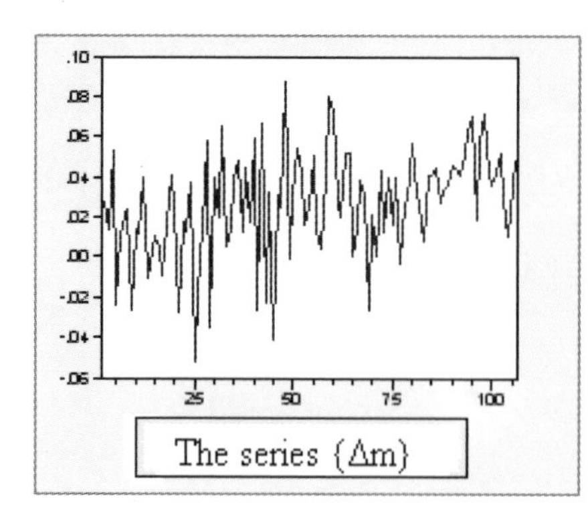

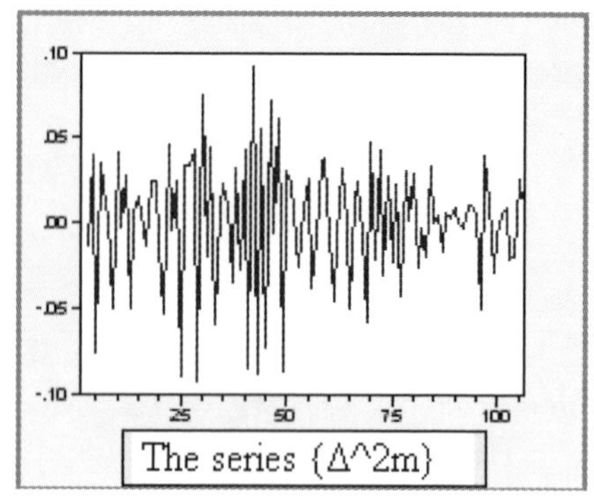

Figure 4. The series $\{\Delta m_i\}$ and $\{\Delta^2 m_i\}$.

Since $F < 8.73$ for $\alpha = 0.01$ (Table 3), the null seen in Eq. (26) is not rejected. Thus, we increase the value of $k (k = 3)$ for the next step. The estimation results, after properly setting the lag-length ($q = 4$), are as follows.

$$\Delta^3 m_i = \underset{(0.004885)}{0.000831} - \underset{(0.495485)}{2.827537} \Delta^2 m_{i-1} - \underset{(0.000077)}{0.000006}\, t_i + \sum_{j=1}^{4} \hat{\beta}_{j+3} \Delta^3 m_{i-j} + \hat{u}_i. \tag{28}$$

No autocorrelation is present. $r_s = -0.164$, $p = 0.108$. Jarque-Bera $= 0.65 (p = 0.723)$. $F(2, 92) = 16.283$. Since the value of F is much greater than 8.73 for $\alpha = 0.01$ (Table 3), the null in Eq. (26) is rejected. Hence, the order of integration of the series $\{m_i\}$ is $k - 1 = 2$, i.e., $\{m_i\} \sim I(2)$. This is also verified from Fig. 4, where $\{\Delta m_i\}$ and $\{\Delta^2 m_i\}$ are plotted.

Even from Fig. 4, one can easily see that the series $\{\Delta m_i\}$ is not stationary. Rather, it seems to be trend-stationary. According to the previous findings, it can be also verified from this figure that $\{\Delta^2 \mathrm{m}_i\}$ is stationary.

It should be noted that in some exceptional or problematic cases, the F-statistic to test Eq. (26), may have a value marginally greater than the corresponding critical value in Table 3, even for $\alpha = 0.01$. In such cases, we propose to increase the value of k and proceed to the next stage.

8. Conclusion

We have derived a simplified method for computing co-integrating vectors, using the SVD. In many applications, we found that this technique produces better results, compared to the ones obtained by applying the ML method.

Due to the properties of SVD, we can easily and directly include in the co-integrating vectors, as many deterministic components as we want to. Besides, we found that the errors corresponding to the finally selected co-integration vector, are characterized by the minimum variance property. It should be noted that the procedure proposed here is not mentioned in the relevant literature.

Further, we analyzed the stages of a relevant technique to obtain reliable results regarding the order of integration of a given series provided that it is DSS. We also demonstrated the properties of the so-called near integrated series, which can be traced by applying this technique. It should be emphasized that with this kind of series (NI), we show that the conventional methods for unit root test may produce faulty results.

References

Banerjee, A, JJ Dolado, JW Galbraith and DF Hendry (1993). *Co-integration, Error Correction, and the Econometric Analysis of Non-Stationary Data.* New York: Oxford University Press.

Dickey, DA and WA Fuller (1981). Likelihood ratio statistics for autoregressive time series with a unit root. *Econometrica* **49**, 1057–1072.

Dickey, DA, DW Jansen and DL Thornton (1994). A primer on cointegration with an application to money and income. In *Cointegration for the Applied Economist*, BB Rao (ed.), UK: St. Martin's Press, 9–45.

Enders, W (1995). *Applied Econometrics Time Series.* New York: John Wiley.

Harris, R (1995). *Using Cointegration Analysis in Econometric Modelling.* London: Prentice Hall.

Holden, D and R Perman (1994). Unit roots and cointegration for the economist. In *Cointegration for the Applied Economist*, BB Rao (ed.), UK: St. Martin's Press, 47–112.

Johansen, S and K Juselius (1990). Maximum likelihood estimation and inference on cointegration with application to the demand for money. *Oxford Bulletin of Economics and Statistics* **52**, 169–210.

Johnston, J and J DiNardo (1997). *Econometric Models.* New York: McGraw-Hill International.

Kleibergen, F and V Dijk (1994). Direct cointegration testing in error correction models. *Journal of Econometrics* **63**, 61–103.

Lazaridis, A and D Basu (1983). Stochastic optimal control by pseudo-inverse. *Review of Economics and Statistics* **65**(2), 347–350.

Lazaridis, A (1986). A note regarding the problem of perfect multicollinearity. *Quantity and Quality* **120**, 297–306.

McKinnon, J (1991). Critical values for co-integration tests. In *Long-Run Economic Relationships*, RF Engle and CWJ Granger (eds.), New York: Oxford University Press.

Perron, P and T Vogelsang (1992a). Nonstationarity and level shifts with an application to purchasing power parity. *Journal of Business and Economic Statistics* **10**, 301–320.

Perron, P and T Vogelsang (1992b). Testing for a unit root in a time series with a changing mean: Corrections and extensions. *Journal of Business and Economic Statistics* **10**, 467–470.

Sims, CA (1980). Macroeconomics and reality. *Econometrica* **48**, 1–48.

Topic 2

Time Varying Responses of Output to Monetary and Fiscal Policy

Dipak R. Basu

Nagasaki University, Nagasaki, Japan

Alexis Lazaridis

Aristotle University of Thessalonikki, Greece

1. Introduction

There are large volumes of literatures on the effectiveness of monetary and fiscal policies and on lags in the effects of monetary policies (Friedman and Schwartz, 1963; Hamburger, 1971; Meyer, 1967; Tobin, 1970). Although the lengths of lags are important in determining the role of monetary-fiscal policies, the relationship between money and income vary over time due to changes in the lag structure. Mayer (1967), Poole (1975) and Warburton (1971) attempted to analyze the empirical variability of the lag structure by analyzing the turning points in general business activity. Sargent and Wallace (1973) as well as Lucas (1972) have drawn attention to the role of time-dependant response coefficients to changes in stabilization policies. Cargill and Meyer (1977; 1978) have estimated the time-varying relationship between national income and monetary-fiscal policies. The results then indicated existences of time variations and exclusions of time variations of the coefficients, lead to exclusions of prior information inherent in the models from the estimation process. Blanchard and Perotti (2002) as well as Smets and Wouters (2003) have obtained similar characteristics of the monetary and fiscal policy.

The existence of time dependency of effects of monetary fiscal policies, reflects considerable doubts on the policy prescription based on constant coefficient estimates. However, more reasonable results can be obtained if we try to estimate dynamics and movements of these relationships over time. For this purpose, adaptive control systems with time-varying parameters (Astrom and Wittenmark, 1995; Basu and Lazaridis, 1986; Brannas and Westlund, 1980; Nicolao, 1992; Radenkovic and Michel, 1992) provide a fruitful approach. There are alternative estimation methods, e.g., rational expectation modeling, VAR approaches etc., to estimate time-varying systems. These methods, however, are mainly descriptive, i.e., cannot have immediate application in policy planning. The purpose of this paper is to explore this possibility in terms of a planning model where adaptive control rules will be implemented so that we can derive time-varying reduced form coefficients from the original structural model to demonstrate the dynamics of monetary-fiscal policies.

The model follows the basic theoretical ideas of monetary approach to balance of payments and structural adjustments (Berdell, 1995; Humphrey, 1981; 1993; Khan and Montiel, 1989). In Section 2, the method of adaptive optimization and the updating method of the time-varying model are analyzed. In Section 3, the model and its estimates are described. This includes some necessary adjustments for a developing country. In Section 4, the results of the time-varying impacts of monetary-fiscal policies are analyzed.

2. The Method of Adaptive Optimization

Suppose a dynamic econometric model can be converted to an equivalent first order dynamic system of the form

$$\tilde{x}_i = \tilde{A}\tilde{x}_{i-1} + \tilde{C}\tilde{u}_i + \tilde{D}\tilde{z}_i + \tilde{e}_i \tag{1}$$

where $\tilde{x}_i$ is the vector of endogenous variables, $\tilde{u}_i$ is the vector of control variables, $\tilde{z}_i$ is the vector of exogenous variables, and $\tilde{e}_i$ is the vector of noises, which are assumed to be white Gaussian and $\tilde{A}$, $\tilde{C}$, and $\tilde{D}$ are coefficient matrices of proper dimensions. It should be noted that a certain element of $\tilde{z}_i$ is 1 and corresponds to the constant terms. The parameters of the above system are assumed to be random.

Shifting to period $i + 1$, we can write

$$\tilde{x}_{i+1} = \tilde{A}\tilde{x}_i + \tilde{C}\tilde{u}_{i+1} + \tilde{D}\tilde{z}_{i+1} + \tilde{e}_{i+1}. \tag{2}$$

Now, we define the following augmented vectors and matrices.

$$x_i = \begin{bmatrix} \tilde{x}_i \\ \tilde{u}_i \end{bmatrix}, \quad x_{i+1} = \begin{bmatrix} \tilde{x}_{i+1} \\ \tilde{u}_{i+1} \end{bmatrix}, \quad e_{i+1} = \begin{bmatrix} \tilde{e}_{i+1} \\ 0 \end{bmatrix},$$

$$A = \begin{bmatrix} \tilde{A} & 0 \\ 0 & 0 \end{bmatrix}, \quad C = \begin{bmatrix} \tilde{C} \\ I \end{bmatrix}, \quad D = \begin{bmatrix} \tilde{D} \\ 0 \end{bmatrix}.$$

Hence (2) can be written as

$$x_{i+1} = Ax_i + C\tilde{u}_{i+1} + D\tilde{z}_{i+1} + e_{i+1} \tag{3}$$

Using the linear advance operator L, such that $L^k y_i = y_{i+k}$ and defining the vectors u, z and ε from

$$u_i = L\tilde{u}_i$$
$$z_i = L\tilde{z}_i$$
$$\varepsilon_i = Le_i$$

then Eq. (3) can take the form

$$x_{i+1} = Ax_i + Cu_i + Dz_i + \varepsilon_i \tag{4}$$

which is a typical linear control system.

We can formulate an optimal control problem of the general form

$$\min J = \frac{1}{2}\|x_T - \hat{x}_T\|^2_{Q_T} + \frac{1}{2}\sum_{i=1}^{T-1}\|x_i - \hat{x}_i\|^2_{Q_i} \tag{5}$$

subject to the system transition equation shown in Eq. (4).

It is noted that T indicates the terminal time of the control period, $\{Q\}$ is the sequence of weighing matrices and $\hat{x}_i (i = 1, 2, \ldots, T)$ is the desired state and control trajectory, according to our formulation.

The solution to this problem can be obtained according to the minimization principle by solving the Ricatti-type equations (Astrom and Wittenmark, 1995).

$$K_T = Q_T \tag{6}$$

$$\Lambda_i = -(E_i C' K_{i+1} C)^{-1}(E_i C' K_{i+1} A) \tag{7}$$

$$K_i = E_i A' K_{i+1} A + \Lambda_i'(E_i C' K_{i-1} A) + Q_i \tag{8}$$

$$h_T = -Q_T \hat{x}_T \tag{9}$$

$$\begin{aligned} h_i = {} & \Lambda_i(E_i C' K_{i+1} D) z_i + \Lambda_i(E_i C') h_{i+1} \\ & + (E_i A' K_{i+1} D) z_i + (E_i A') h_{i+1} - Q_i \hat{x}_i \end{aligned} \tag{10}$$

$$g_i = -(E_i C' K_{i+1} C)^{-1}[(E_i C' K_{i+1} D) z_i + (E_i C') h_{i+1}] \tag{11}$$

$$x_i^* = [E_i A + (E_i C)\,\Lambda_i]\,x_i^* + (E_i C)\,g_i + (E_i D) z_i \tag{12}$$

$$u_i^* = \Lambda_i x_i^* + g_i \tag{13}$$

where $u_i^*(i = 0, 1, \ldots, T-1)$, the optimal control sequence and x_{i+1}^*, the corresponding state trajectory, which constitutes the solution to the stated optimal control problem.

In the above equations, Λ_i is the matrix of feedback coefficients and g_i is the vector of intercepts. The notation E_i denotes the conditional expectations, given all information up to the period i. Expressions like $E_iC'K_{i+1}C$, $E_iC'K_{i+1}A$, $E_iC'K_{i+1}D$ are evaluated, taking into account the reduced form coefficients of the econometric model and their co-variance matrix, which are to be updated continuously, along with the implementation of the control rules. These rules should be re-adjusted according to "passive-learning" methods, where the joint densities of matrices A, C, and D are assumed to remain constant over the control period.

2.1. *Updating Method of Reduced-Form Coefficients and Their Co-Variance Matrices*

Suppose we have a simultaneous-equation system of the form

$$XB' + U\Gamma' = R \tag{14}$$

where X is the matrix of endogenous variable defined on $E^N \times E^n$ and B is the matrix of structural coefficients, which refer to the endogenous variables and is defined on $E^n \times E^n$. U is the matrix of explanatory variables defined on $E^N \times E^g$ and Γ is the matrix of the structural coefficients, which refer to the explanatory variables, defined on $E^N \times E^g$. R is the matrix of noises defined on $E^N \times E^n$. The reduced form coefficients matrix Π is then defined from:

$$\Pi = -B^{-1}\Gamma. \tag{14a}$$

Goldberger *et al.* (1961) have shown that the asymptotic co-variance matrix, say Ω of the vector $\hat{\pi}$, which consists of the g column of matrix $\hat{\Pi}$ can be approximated by

$$\tilde{\Omega} = \left[\begin{bmatrix}\hat{\Pi} \\ I_g\end{bmatrix} \otimes (\hat{B}')^{-1}\right]' F \left[\begin{bmatrix}\hat{\Pi} \\ I_g\end{bmatrix} \otimes (\hat{B}')^{-1}\right] \tag{15}$$

where $\otimes$ denotes the Kroneker product, $\hat{\Pi}$ and $\hat{B}$ are the estimated coefficients by standard econometric techniques and F denotes the asymptotic co-variance matrix of the $n + g$ columns of $(\hat{B}\hat{\Gamma})$, which assumed to be consistent and asymptotically unbiased estimate of $(B\Gamma)$.

Combining Eqs. (14) and (14a) we have

$$BX' = -\Gamma U' + R' \Rightarrow X' = -B^{-1}\Gamma U' + B^{-1}R'$$
$$\Rightarrow X' = \Pi U' + W' \qquad (16)$$

where $W' = B^{-1}R'$

Denoting the ith column of matrix X' by x_i and the ith column of matrix W' by w_i, we can write

$$x_i = \begin{bmatrix} u_{1i} & 0 & \cdots & 0 & u_{2i} & 0 & \cdots & 0 & u_{gi} & 0 & \cdots & 0 \\ 0 & u_{1i} & \cdots & 0 & 0 & u_{2i} & \cdots & 0 & 0 & u_{gi} & \cdots & 0 \\ \cdot & \cdot & & \cdot & \cdot & \cdot & & \cdot & \cdot & \cdot & & \cdot \\ \cdot & \cdot & & \cdot & \cdot & \cdot & & \cdot & \cdot & \cdot & & \cdot \\ \cdot & \cdot & & \cdot & \cdot & \cdot & & \cdot & \cdot & \cdot & & \cdot \\ 0 & 0 & \cdots & u_{1i} & 0 & 0 & \cdots & u_{2i} & \cdots & 0 & \ldots & u_{gi} \end{bmatrix} \pi + w_i \qquad (17)$$

where u_{ij} is the element of the jth column and ith row of matrix U. The vector $\pi \in E^{ng}$, as mentioned earlier, consists of the g column of matrix Π.

Equation (17) can be written in a compact form, as

$$x_i = \mathrm{H}_i\pi + w_i, \quad i = 1, 2, \ldots, N \qquad (17a)$$

where $x_i \in E^n$, $w_i \in E^n$ and the observation matrix H_i is defined on $E^n \times E^{ng}$.

In a time-invariant econometric model, the coefficients vector π is assumed random with constant expectation overtime, so that

$$\pi_{i+1} = \pi_i, \quad \text{for all } i. \qquad (18)$$

In a time-varying and stochastic model we can have

$$\pi_{i+1} = \pi_i + \varepsilon_i \qquad (18a)$$

where $\varepsilon_i \in E^{ng}$ is the noise.

Based on these, we can re-write Eq. (17a) as

$$x_{i+1} = H_{i+1}\pi_{i+1} + w_{i+1} \quad i = 0, 1, \ldots, N-1. \qquad (19)$$

We make the following assumptions.

(a) The vector x_{i+1} and matrix H_{i+1} can he measured exactly for all i.
(b) The noises ε_i and w_{i+1} are independent discrete white noises with known statistics, i.e.,

$$E(\varepsilon_i) = 0; \quad E(w_{i+1}) = 0$$
$$E(\varepsilon_i w'_{i+1}) = 0$$

$$E(c_i\varepsilon_i') = Q_1\delta_{ij}, \quad \text{where } \delta_{ij} \text{ is the Kronecker delta, and}$$
$$E(w_i w_i') = Q_2\delta_{ij}.$$

The above co-variance matrices, assumed to be positive definite.

(c) The state vector is normally distributed with a finite co-variance matrix.

(d) Regarding Eqs. (18a) and (19), the Jacobians of the transformation of ε_i into π_{i+1} and of w_{i+1} into x_{i+1} are unities. Hence, the corresponding conditional probability densities are:

$$p(\pi_{i+1}|\pi_i) = p(\varepsilon_i)$$
$$p(x_{i+1}|\pi_{i+1}) = p(w_{i+1}).$$

Under the above assumptions and given Eqs. (18a) and (19), the problem set is to evaluate

$$E(\pi_{i+1}|x^{i+1}) = \pi_{i+1}^*$$

and

$$\text{cov}\,(\pi_{i+1}|x^{i+1}) = S_{i+1} \quad \text{(the error co-variance matrix)}$$

where $x^{i+1} = x_1, x_2, x_3, \ldots, x_{i+1}$.

The solution to this problem (Basu and Lazaridis, 1986; Lazaridis, 1980) is given by the following set of recursive equations, as it is briefly shown in Appendix A.

$$\pi_{i+1}^* = \pi_i^* + K_{i+1}(x_{i+1} - H_{i+1}\pi_i^*) \tag{20}$$
$$K_{i+1} = S_{i+1}H_{i+1}'Q_2^{-1} \tag{21}$$
$$S_{i+1}^{-1} = P_{i+1}^{-1} + H_{i+1}'Q_2^{-1}H_{i+1} \tag{22}$$
$$P_{i+1}^{-1} = (Q_1 + S_i)^{-1}. \tag{23}$$

The recursive process is initiated by regarding K_0 and H_0 as null matrices and computing π_0^* and S_0 from

$$\pi_0^* = \hat{\pi} \quad \text{i.e., the reduced form coefficients (columns of matrix } \hat{\Pi})$$
$$S_0 = P_0 = \tilde{\Omega}.$$

The reduced form coefficients, along with their co-variance matrices, can be updated by this recursive process and at each stage the set of Riccati equations should be updated accordingly so that adaptive control rules may be derived.

Once we estimate the model (described in the next section), using the full information maximum likelihood (FIML) method, we can obtain both the structural model and the probability density functions along with all associated matrices mentioned above. We first convert the structural econometric model to a *state-variable* form according to Eq. (1). Once we specify the targets for the state and control variables, the objective function is to be minimized, the weights attached to each state and control variables, then we can calculate the results of the optimization process for the entire period using Eqs. (6)–(13). Thereafter, we can update all probability density functions and all other associated matrices using Eqs. (20)–(23). These will effectively update the coefficients of the model in its *state-variable* form. We can repeat the optimization process over and over as we update the model, its associated matrices, probability density functions and use these as new information. When the process will converge, i.e., the difference between the old and new estimates are insignificant according to some preassigned criteria, we can obtain the final updated coefficients of the model along with the results of the optimization process. The dynamics of the time-varying coefficients (in terms of response-multipliers) are described later in Section 4. This mechanism is a great improvement compared to the fixed-coefficient model as we can incorporate in our calculation, changing information sets derived from the implementation of the optimization process.

3. Monetary-Fiscal Policy Model

We describe below a monetary-fiscal policy model, which was estimated with data from the Indian economy. The model accepts the definition of balance of payments and money stock according to monetary approach of balance of payments (Khan, 1976; Khan and Montiel, 1989; 1990). The decision to choose this particular model is influenced by the fact that it is commonly known as the fund-bank adjustment policy model, used extensively by the World Bank and the International Monetary Fund (IMF).

In the version adopted for this paper, there is no explicit investment or consumption function, because the domestic absorption can reflect the combined response of both private and public investment and consumption, to the planned target for national income set by the planning authority and to various market forces reflected in the money market interest rate and exchange rate. The "New Cambridge" model of the UK economy (Cripps and Godley, 1976; Godley and Lavoie, 2002; 2007) has postulated a similar combined consumption-investment function for the UK as well.

The model was estimated by applying FIML method, using the data from the Indian economy for the period 1951–1996. The estimated parameters were then used as the initial starting point for the stochastic control model. The estimated econometric model can be transformed into a state variable form (Basu and Lazaridis, 1986), in order to formulate the optimal control problem, i.e., to minimize the quadratic objective function, subject to Eq. (4), which is the system transition equation. As already mentioned, the recursive estimation process of the time-varying response-multipliers is presented in brief in Appendix A.

3.1. *Absorption Function and National Income*

Domestic absorption reflects the behavior of both the private and public sector regarding consumption and investment too. Domestic real adsorption is influenced by real national income, market interest rate and foreign exchange rate. We assume a linear relationship.

$$(A/P)_t = a_0 + a_1(Y/P)_t - a_2(IR)_t - a_3 EXR_t, \quad t = \text{time period.}$$

The relation between the national income and adsorption can be defined as follows:

$$Y_t = A_t + TY_t + \Delta R_t - G_t + GBS_t - LR_t$$

where A is the value of domestic absorption, P is the price level, Y is the national income, IR is the market interest rates, EXR is the exchange rate, TY is the government tax revenue, G is the public consumption, GBS is the government bond sales, LR is the net lending by the central government to the states (which is not part of the planned public expenditure) and ΔR is the changes in the foreign exchange reserve reflecting the behavior of the foreign trade sector.

The government budget deficit (BD_t) is defined by the following equation

$$BD_t = (G_t + LR_t + PF_t) - (TY_t + GBS_t + AF_t + BF_t)$$

where PF is the foreign payments due to existing foreign debts, which may include both amortization and interest payments, AF is the foreign assistance which is an insignificant feature, *FB* is the total foreign borrowing, assuming only the government can borrow from foreign sources.

We assume *AF* and *LR* as exogenous, whereas *FB*, *G*, and *TY* as policy instruments. However, *PF* depends on the level of existing foreign debt and the world interest rate, although a sizable part of the foreign borrowing can be at a concessional rate

$$PF_t = a_4 + a_5 \sum_{r=-20}^{t} FB_r + a_6 \left(\frac{WIR}{EXR}\right)_t .$$

Government bond sales (*GBS*) depends on its attractiveness reflected on the interest rate (*IR*), on the ability of the domestic economy to absorb (A) on the requirements of the government (G) and on the alternative sources of finances reflected on the tax revenue (*TY*) and on government's borrowing from the central bank i.e., the net domestic asset (NDA) creation by the central bank.

$$GBS_t = a_7 + a_8 A_t + a_9 IR_t + a_{10} G_t - a_{11} TY_t - a_{12} NDA_t$$

3.2. *Monetary Sector*

We assume the flow equilibrium in the money market, i.e.,

$$\Delta MD_t = \Delta M_t = \Delta MS_t$$

where *MD* is the money demand and *MS* is the money supply. The stock of money supply depends on the stock of high powered money and the money-multiplier, as follows:

$$MS_t = [(1 + CD)/(CD + RR)]_t (\Delta R + NDA)_t.$$

$(\Delta R + NDA)$ reflect the stock of high-powered money and the expression within the square bracket is the money-multiplier, which depends on credit to deposit ratio of the commercial banking sector (*CD*) and the reserve to deposit liabilities in the commercial banking sector (*RR*). Whereas *NDA* is an instrument, ΔR depends on the foreign trade sector. However, the government can influence *CD* and *RR* to control the money supply. *RR*, which is the actual reserve ratio, depends on the demand for loans created by private sectors and the commercial bank's willingness to lend. We assume that the desired reserve ratio RR_t^* is a function of national income and market interest rate i.e.,

$$RR_t^* = a_{13} + a_{14}Y_t + a_{15}IR_t^*.$$

The commercial banks may adjust their actual reserve ratio to the desired reserve ratio with a lag.

$$\Delta RR_t = \alpha(RR_t^* - RR_{t-1}) \quad \text{where } 0 < \alpha < 1.$$

Thus, we can write

$$RR_t = \alpha a_{13} + \alpha a_{14}Y_t + \alpha a_{15}IR_t + (l - \alpha)RR_{t-1}.$$

The ratio of credit to deposit liabilities with the commercial bank system is affected by the opportunity cost of holding currency, as measured by the market interest rate and national income, representing the domestic economic activity. Khan (1976) has postulated that the effect of national income should be negative because "individuals and corporations tend to become more efficient in their management of cash balances as their income rises". However, a different logic may emerge to a developing country where the use of banks as an institution is not widespread, particularly among the labor force. If there is an expansion in economic activity, the entrepreneurs will have to maintain a huge cash-balance and run down deposits simply to pay various dues, as most payments would have to be made in cash. It is possible that corporations would be more efficient, after some initial adjustment period. We, therefore, expect the sign of the coefficient for the current national income to be positive and that for the lagged national income to be negative.

$$CD_t = a_{16} + a_{17}IR_t + a_{18}Y_t - a_{19}Y_{t-1}.$$

The demand for money is assumed to be a function of the market interest rate and the national income.

$$MD_t = a_{20} - a_{21}IR_t + a_{22}Y_t.$$

3.3. *Price and Interest Rate*

The market rate of interest (*IR*) is determined by the supply of money, national income, and the central bank discount rate.

$$IR_t = a_{23} - a_{29}(MS_t) + a_{29}(Y_t) + a_{26}(CI_t).$$

The domestic price level depends on domestic economic activity (particularly changes in the agricultural sector) and the import cost (*IMC*).

The import cost in turn depends on the exchange rate (*EXR*) and the world price of imported goods (*WPM*). We assume the desired price level (P^*) is represented by the following equation:

$$P_t^* = a_{27} - a_{28}(A_t) + a_{29}(IMC_t).$$

The desired price level reflects private sector's reaction to their expected domestic absorption of the expected import cost. Suppose the actual price will move according to the difference between the desired price in period t and the actual price level in the previous period

$$\Delta Pt = \beta(P_t^* - P_{t-1}); \quad 0 < \beta < 1.$$

Thus, we get

$$P_t = \beta a_{27} - \beta a_{27}(A_t) + \beta a_{29}(IMC_t) + (1 - \beta)P_{t-1}.$$

Import cost (*IMC*) is represented by the following equation time-varying responses of output to monetary and fiscal policy

$$IMC_t = a_{30} - a_{31}(EXR_t) + a_{32}(WPM_t)$$

Exchange rate (*EXR*) can be an instrument variable whereas *WPM* is an exogenous variable.

3.4. *Balance of Payments*

The balance of payments (ΔR) is equal to the changes in the stock of international reserve i.e.,

$$\Delta R_t = X_t - IM_t + K_t + PFT_t + FB_t - PF_t + AF_t$$

where X is the value of exports, *IM* is the value of imports, K is the foreign capital inflows, *PFT* is the private sectors transactions, *FB* is the foreign borrowing, *PF* is the foreign payments by the central bank and *AF* is the foreign aid and grants; where X, *PFT*, K and *AF* are exogenous, import *IM* is determined by the national income, and the import cost i.e.,

$$IM_t = a_{33} + a_{34}Y_t - a_{35}(IMC_t).$$

The above analytical structure was estimated using expected values of each variable, with expectations being adaptive. The estimated parameters were used as the initial starting point for the stochastic control model developed. Estimated model is presented in Appendix B.

3.5. *Stability of the Model*

Characteristic roots (real) of the system transition matrix

$$
\begin{array}{r}
-0.0000008 \\
-0.0000008 \\
-0.1213610 \\
0.2442519 \\
0.3087129 \\
0.5123629 \\
0.7824260 \\
0.0000001 \\
0.0000001
\end{array}
$$

All the roots are real and they are less than unity, so the system is stable. [In the equivalent control system, the rank of the controllability matrix is the same as the dimension of the reduced state vector so that the system is controllable and observable, i.e., the system parameters can be identified.]

We can, however, transform our model to the following form:

$$Y_t = \rho Y_{t-1} + e_t, \quad t = 1, 2, \ldots, n \tag{24}$$

where ρ is a real number and (e_t) is a sequence of normally distributed random variables with mean zero and variance σ_t^2. Box and Pierce (1970) suggested the following test statistic

$$Q_m = n \sum_{k=1}^{m} r_k^2 \tag{25}$$

where

$$r_k = \sum_{t=k+1}^{n} \hat{e}_t \hat{e}_{t-k} \Bigg/ \left(\sum_{t=1}^{n} \hat{e}_t^2 \right) \tag{26}$$

$n =$ the number of observations, $m = n - k$, where $k =$ the number of parameters estimated, and $\hat{e}_t$ are the residuals from the fitted model.

If (Y_t) satisfies the system, then under the null hypothesis, Q_m is distributed as a chi-squared random variable with m degrees of freedom. The null hypothesis is that $\rho = 1$ where $\hat{e}_t = Y_t - Y_{t-1}$ and thus $k = 0$.

The estimated value of Q_m is 3.85, where the null hypothesis is rejected at 0.05 significance level. So, we accept the alternative hypothesis that if $\rho < 1$, therefore the system is stable.

Das and Cristi (1990) have analyzed in detail the condition for the stability and robustness of the time-varying stochastic optimal control system. The condition is that the dynamic response-multipliers of the model should have slow time-variations. As demonstrated in Tables 1–7, the response-multipliers satisfy conditions of slow time-variations (Das and Cristi, 1990; Tsakalis and Ioannou, 1990).

Table 1. Response-multiplier[a] and endogenous variable.

Exogenous variable	*Y*	*GBS*	*P*	*BD*	*CD*	*MS*	*IR*
EXR	−0.02283	−0.00027	0.05593	0.00121	−0.00018	−0.00576	−0.00004
G	0.03178	0.00095	0.13965	0.04073	0.00041	0.01699	0.00108
TY	−0.03116	−0.00426	−0.15022	0.04091	−0.00310	−0.03637	−0.01231
CI	−0.00265	0.04953	−0.00884	0.00550	−0.00210	−0.00625	0.06599

[a]*Note*: This response-multiplier refers to the starting point of the planning; $\lambda^2 = 563.09$, indicating overall goodness of fit in a FIML estimation.

Table 2. Response-multiplier period 1.[a]

Endogenous	*Y*	*GBS*	*P*	*BD*	*CD*	*MS*	*IR*
EXR	−0.02254	−0.00092	0.05458	0.00111	−0.00020	−0.00543	−0.00092
G	0.03572	0.00403	0.14849	0.03987	0.00500	0.01466	0.00330
TY	−0.03590	−0.01017	−0.15549	0.02791	−0.00399	−0.02682	−0.01996
CI	0.00226	0.04592	−0.00606	−0.00458	−0.00169	−0.00516	0.06120

[a]*Note*: Period 1 refers to the first period of the planning horizon,

Table 3. Response-multiplier period 2.

Endogenous	*Y*	*GBS*	*P*	*BD*	*CD*	*MS*	*IR*
EXR	−0.02209	−0.00096	0.05336	0.00125	−0.00013	−0.00570	−0.00097
G	0.03293	0.00568	0.15603	0.03500	0.00091	0.02882	0.00599
TY	−0.02961	−0.01240	−0.15688	0.02792	−0.00246	−0.04758	−0.02324
CI	0.00273	0.94546	−0.00387	−0.00488	−0.00181	−0.00668	0.06057

Table 4. Response-multiplier period 3.

Endogenous	*Y*	*GBS*	*P*	*BD*	*CD*	*MS*	*IR*
EXR	−0.02199	−0.00069	0.05269	0.00133	−0.00042	−0.00573	−0.00062
G	0.03001	0.00499	0.16536	0.02924	0.05923	0.04217	0.00570
TY	−0.02470	−0.01149	−0.16281	0.02472	−0.07031	−0.06325	−0.02220
CI	0.00215	0.04676	−0.00252	−0.00399	−0.00465	−0.00671	0.06228

Table 5. Response-multiplier period 5.

Endogenous	*Y*	*GBS*	*P*	*BD*	*CD*	*MS*	*IR*
EXR	−0.02133	−0.00088	0.05190	0.00152	−0.00201	−0.00595	−0.00089
G	0.02058	0.00558	0.18078	0.01160	0.50890	0.07248	0.00530
TY	−0.01676	−0.00410	−0.16655	0.00695	−0.55026	−0.07779	−0.01013
CI	0.00268	0.04465	−0.00267	−0.00403	−0.01967	−0.00726	0.05953

Table 6. Response-multiplier period 7.

Endogenous	*Y*	*GBS*	*P*	*BD*	*CD*	*MS*	*IR*
EXR	−0.02067	−0.00116	0.04987	0.00135	−0.00104	−0.00558	−0.00129
G	0.00733	0.01033	0.16455	0.00048	0.57357	0.10760	0.01206
TY	−0.00402	−0.00345	−0.14832	−0.01434	−0.58067	−0.10904	−0.00869
CI	0.00200	0.04174	−0.00236	−0.00293	−0.02186	−0.00551	0.05566

Table 7. Response of output to monetary and fiscal policies.

Periods	1	2	3	5	7
dY/dCI	−0.00226	−0.00273	−0.00215	−0.00268	−0.00200
$dY/dEXR$	−0.02254	−0.02209	−0.02188	−0.02133	−0.02067
dP/dG	0.14849	0.15603	0.16536	0.18078	0.16455
dY/dG	0.03572	0.03203	0.03001	0.02058	0.00733
dP/dTY	−0.15549	−0.15688	−0.16281	−0.16655	−0.14832
dY/dTY	−0.03590	−0.02691	−0.02470	−0.01676	−0.00402
$dGBS/dG$	0.00403	0.00568	0.00499	0.00558	0.01033

4. Policy Analysis

The dynamics of the monetary and fiscal policy are analyzed in terms of the analysis of the dynamic response-multipliers and their relationship with the monetary-fiscal policy regimes. The response-multiplier of the system will move over time within an adaptive control framework (Tables 1–6). The movements of the coefficients response of response-multipliers in a monetary policy framework are described below. It is essential that for the stability of the control system, these dynamics of the response-multiplier should be slow (Das and Cristi, 1990; Tsakalis and Ioannou, 1990).

Analysis of the response-multipliers shows that devaluation would have negative effect on the national income, devaluation would also have negative effect on the government bond sales, CD, interest rate, and money demand. This is due to the fact that Indian exports may not be that elastic in response to devaluation, whereas devaluation will reduce import abilities significantly. As a result, national income and domestic activity would have negative effects, which would depress the activity of the private sector. Because of this CD will decline.

Due to the downturn of the economic activity, there will be less demand for government bonds. Devaluations also can have an inflationary effect due to increased import costs. Government expenditure, on the other hand, will have positive effects on every variable. This implies that increased public expenditure will stimulate the national income and the private sector despite increased interest rate. However, the increased public expenditure will have inflationary impacts on the economy at the same time.

Increased tax revenue will depress national income, as it will reduce government bond sales. Lower level of national income will reduce money supply and the private sector's activity, which will be reflected in the reduced currency to deposit ratio and the reduced level of money demand. Reduced level of national income, in this case, will have reduced price level but budget deficit will go up as well because of the reduced demand for government bonds. Increased rate of central bank's discount rate will depress national income and general price level. At the same time, private sector's activity will be reduced, as reflected in the *CD* and money demand, due to increased rate of interest. Although government bond sales will go up, budget deficit will be increased due to reduced level of economic activity.

The effect of the central bank discount rate (*CI*) on interest rate varies from 0.061 in period 1 to 0.055 in period 7. This result is primarily due to the fact that the influence of *G* (public expenditure) on IR has increased from 0.003 in period 1 to 0.012 in period 7. If the public expenditure goes up, there will be direct pressure on the central bank to provide loans to the government. The central bank can put pressure on the commercial banks to extend some loans in public sector and less to the private sector, and the *CD* of the commercial banks will be reduced as a result (at the same time reserve ratio will be increased). While *CD* goes up in response to government expenditure (G), the *CI* will have a similar effect on the currency to deposit ratio.

Effects of the *CD* on prices declines rapidly, whereas the effect of the exchange rate on price level is most prominent and does not vary much. It demonstrates that although during the initial phase of planning, tight credit policy may influence prices, in the longer run structural factors, public expenditure, and import costs reflected through the exchange rate will affect price levels more significantly. The effect of the *CI* on the money supply (*MS*) will be reduced from −0.005 in period 1 to −0.007 in period 6, and it will slightly increase in period 7.

The impacts of the public expenditure and tax revenues on the two most important variables, *GNP* (Y) and price level (P) can be analyzed as follows. The increased government expenditure (G) will increase price level and the impact will be intensified. The impact of the government expenditure on the *GNP* will be reduced, which is consistent with the reduced impacts of the tax revenues on the *GNP* over the planning period. Tax revenue will have negative impacts on the *GNP* as well as on the price level, and the impacts will decline over time. The negative impacts of the tax revenue on the *GNP* is partly explained by the negative effects of tax revenue on the currency to deposit ratios of the commercial banks, the main indicator for the private sector activities. The government expenditure also have positive impacts on the government bond sales and the impact will increase over time due to the increasing difficulties of raising taxes, which will have negative impacts on the growth prospects.

The effect of the exchange rate declines gradually, whereas the effect of the interest rate is cyclical. This is due to the cyclical pattern of the central discount rate in the optimum path. It is quite obvious that the impacts of fiscal and monetary policies on output will fluctuate over time depending on the direction of change of these policies. The variations of the effects are not unsystematic, as feared by Friedman and Schwartz (1963). However, here we are analyzing an optimum path not the historical path.

5. Conclusion

The importance of time-varying coefficients was recognized in statistical literature long ago (Davis, 1941), where frequency domain approach of the time series analysis was utilized (Mayer, 1972). Several authors since then have used varying parameter regression analysis (Cooley and Prescott, 1973; Farley *et al.*, 1975). In the above analysis, adaptive control system was used in a state-space model where the reduced form parameters can move over time.

However, variations over time are slow which indicates any absence of explosive response (Das and Cristi, 1990; Tsakalis and Ioannou, 1990). Cargil and Mayer (1978) also have observed stable movements of the coefficients. Thus the role of the monetary-fiscal policies on the economy is not unsystematic, although it can vary over time.

Results obtained by other researchers showed that the effects of monetary and fiscal policy change over time, and it is important to analyze these changes in order to obtain time-consistent monetary-fiscal policy (De Castro, 2006; Folster and Henrekson, 2001; Muscatelli and Tirelli, 2005). The results obtained by using adaptive control method, showed similar characteristics of the monetary-fiscal policy.

The implication for the public policy is quite obvious. Time-consistent monetary-fiscal policy demands continuous revision, otherwise, the effectiveness of the policy may deteriorate and as a result, the effects of monetary and fiscal policy on major target variables of the economy may deviate from their desired level. Our approach is a systematic way forward to analyze these dynamics of monetary and fiscal policy.

References

Astrom, KJ and K Wittenmark (1995). *Adaptive Control.* Reading, MA: Addison-Wesley.

Basu, D and A Lazaridis (1986). A method of stochastic optimal control by Bayesian filtering techniques. *International Journal of System Sciences* **17**(1), 81–85.

Brannas, K and A Westlund (1980). On the recursive estimation of stochastic and time varying parameters in economic system. In *Optimization Techniques*, K Iracki (ed.), Berlin: Springer Verlag, 265–274.

Berdell, JF (1995). The prevent relevance of Hume's open-economy monetary dynamics. *Economic Journal* **105**(432), September, 1205–1217.

Box, GEP and DA Pierce (1970). Distribution of residual autocorrelation in autoregressive-integrated moving average time series models. *Journal of the American Statistical Association* **65**(3), 1509–1526.

Blanchard, O and R Perotti (2002). An empirical characterization of the dynamic effects of changes in government spending and taxes on output. *Quarterly Journal of Economics*, **117**(4), November, 1329–1368.

Cargil, TF and RA Meyer (1977). Intertemporal stability of the relationship between interest races and prices. *Journal of Finance* **32**, September, 427–448.

Cargil, TF and RA Mayer (1978). The time varying response of income to changes in monetary and fiscal policy. *Review of Economics and Statistics* **LX**(1), February, 1–7.

Cripps, F and WAH Godley (1976). A formal analysis of the Cambridge economic policy group Model. *Economica* **43**, August, 335–348.

Cooley, TF and EC Prescott (1973). An adaptive regression model. *International Economic Review* **l4**, June, 248–256.

Das, M and R Cristi (1990). Robustness of an adaptive pole placement algorithm in the presence of bounded disturbances and slow time variation of parameters. *IEEE Transactions on Automatic Control* **35**(6), June, 762–802.

Davis, HT (1941). *The Analysis of Economic Time Series*. Bloomington: Principia Press.

De Castro, F (2006). Macroeconomic effects of fiscal policy in Spain. *Applied Economics* **38**(8–10), May, 913–924.

Farley, J, M Hinich and T McGuire (1975). Some comparison of tests for a shift in the slopes of a multivariate time series model. *Journal of Econometrics* **3**, August, 297–318.

Friedman, M and A Schwartz (1963). A monetary history of the United States, 1867–1960. Princeton: Princeton University Press.

Folster, S and M Henrekson (2001). Growth effects of government expenditure and taxation in rich countries. *European Economic Review* **45**(8), 1501–1520.

Goldberger, AC, AL Nagar and HS Odeh (1961). The covariance matrices of reduced form coefficients and forecasts for a structural econometric model. *Econometrica* **29**, 556–573.

Godley, W and M Lavoie (2002). Kaleckian models of growth in a coherent stock-flow monetary framework: A Kaldorian view. *Journal of Post Keynesian Economics* **24**(2), 277–312.

Godley, W and M Lavoie (2007). Fiscal policy in a stock-flow consistent (SFC) model. *Journal of Post Keynesian Economics* **30**(1), 79–100.

Hamberger, MJ (1971). The lag in the effect of monetary policy: A survey of recent literature. *Monthly Review-Federal Reserve Bank of New York*, December, 289–297.

Humphrey, TM (1981). Adam Smith and the monetary approach to the balance of payments. *Economic Review, Federal Reserve Bank of Richmond* **67**.

Humphrey, TM (1993). *Money, Ranking and Inflation*. Aldershot: Edward Elgar.

Khan, MS (1976). A monetary model of balance of payments: The case of Venezuela. *Journal of Monetary Economics* **2**, July, 311–332.

Khan, MS and PJ Montiel (1989). Growth oriented adjustment programs. *IMF Staff Papers* **36**, June, 279–306.

Khan, MS and PJ Montiel (1990), A marriage between fund and bank models. *IMF Staff Papers* **37**, March, 187–191.

Lazaridis, A (1980). Application of filtering methods in econometrics. *International Journal of Systems Science* **11**(11), 1315–1325.

Lucas, RE, Jr (1972). Expectation and the neutrality of money. *Journal of Economic Theory* **4**, April, 103–124.

Meyer, T (1967). The lag in effect of monetary policy, some criticisms. *Western Economic Journal* **5**, September.

Mayer, R (1972) Estimating coefficients that change over time. *International Economic Review* **11**, October.

Muscatelli, VA and P Tirelli (2005). Analyzing the interaction of monetary and fiscal policy: does fiscal policy play a valuable role in stabilization. *CESIFO Economic Studies* **51**(4), 1–17.

Nicolao, G (1992). On the time-varying Riccati difference equation of optimal filtering. *SIAM Journal on Control and Optimization* **30**(6), November, 1251–1269.

Poole, W (1975). The relationship of monetary deceleration to business cycle peaks. *Journal of Finance* **30**, June, 697–712.

Radenkovic, M and A Michel (1992). Verification of the self-stabilization mechanism in robust stochastic adaptive control using Lyapunov function arguments. *SIAM Journal on Control and Optimization* **30**(6), November, 1270–1294.

Sargent, TJ and N Wallace (1973). Rational expectations and the theory of economic policy. *Journal of Monetary Economics* **2**, April, 169–183.

Smets, V and R Wouters (2003). An estimated dynamic stochastic general equilibrium model of the Euro area. *Journal of European Economic Association* **1**(5), September, 1123–1175.

Tobin, J (1970). Money and income: Post Hoc Ergo Propter Hoc. *Quarterly Journal of Economics* **84**, May, 301–317.

Tsakalis, K and P Ioannou (1990). A new direct adaptive control scheme for time-varying plant. *IEEE Transactions on Automatic Control* **35**(6) June, 356–369.

Warburton, C (1971). Variability of the lag in the effect of monetary policy 1919–1965. *Western Economic Journal* **9**, June, 1919–1965.

Appendix A

Probability Density Function and Recursive Process for Re-Estimation

Under the assumptions stated in Section 3 and according to Bayes' rule, the conditional probability density function of π_i given x^{i+1} is Gaussian and is given by:

$$p(\pi_{i+1}|x^{i+1}) = \int p(\pi_i|x^{\mathrm{i}})p(\pi_{\mathrm{i}+1}, x_{i+1}|\pi_i, x^i)d\pi_{\mathrm{i}}$$

where

$$p(\pi_{i+1}|x^{\mathrm{i}}) = \text{constant exp}\left(-\frac{1}{2}||\pi_{i+1} - \pi_i^*||^2_{S_i^{-1}}\right)$$

$$\pi_i^* = E(\pi_{\mathrm{i}}|x^i)$$

$$S_i = \text{cov}(\pi_{\mathrm{i}}|x^i) \quad (S_i \text{ assumed to be invertible})$$

$$P(\pi_{i+1}, x_{i+1}|\pi_i, x^i) = \text{constant exp}\left[-\frac{1}{2}\left\|\begin{matrix}\pi_{i+1} - \pi_i \\ x_{i+1} - H_{i+1}\pi_{i+1}\end{matrix}\right\|^2_{C^{-1}}\right]$$

$$C = \begin{bmatrix} Q_1 & 0 \\ 0 & Q_2 \end{bmatrix}, \quad C^{-1} = \begin{bmatrix} Q_1^{-1} & 0 \\ 0 & Q_2^{-1} \end{bmatrix}.$$

Hence

$$p(\pi_{i+1}|x^{i+1}) = \text{constant} \int \exp\left(-\frac{1}{2}\tilde{J}_i\right) d\pi_i$$

where

$$\begin{aligned}
\tilde{J}_i = {} & (\pi_i - \pi_i^*)'(S_i^{-1} + Q_1^{-1})(\pi_i - \pi_i^*) \\
& + (\pi_{i+1} - \pi_i^*)'(Q_1^{-1} + H_{i+1}' Q_2^{-1} H_{i+1})(\pi_{i+1} - \pi_i^*) \\
& - 2(\pi_i - \pi_i^*)' Q_1^{-1}(\pi_i - \pi_i^*) \\
& + (x_{i+1} - H_{i+1}\pi_i^*)' Q_2^{-1}(x_{i+1} - H_{i+1}\pi_i^*) \\
& - 2(\pi_{i+1} - \pi_i^*)' H_{i+1}' Q_2^{-1}(x_{i+1} - H_{i+1}\pi_i^*). \qquad \text{(a)}
\end{aligned}$$

Now, define $G_i^{-1} = (S_i^{-1} + Q_1^{-1})$ and consider the expression

$$\begin{aligned}
\tilde{\tilde{J}}_i = {} & [(\pi_i - \pi_i^*)' - G_i Q_1^{-1}(\pi_{i+1} - \pi_i^*)]' \\
& \times G_i^{-1}[(\pi_i - \pi_i^*)' - G_i Q_1^{-1}(\pi_{i+1} - \pi_i^*)]. \qquad \text{(b)}
\end{aligned}$$

Note that G_i is symmetric, since it is the sum of two (symmetric) co-variance matrices. Expanding Eq. (b) and noting that $G_i G_i^{-1} = I$ we obtain

$$\begin{aligned}
\tilde{\tilde{J}}_i = {} & (\pi_i - \pi_i^*)' G_i^{-1}(\pi_{i+1} - \pi_i^*) - 2(\pi_i - \pi_i^*)' Q_1^{-1}(\pi_{i+1} - \pi_i^*) \\
& + (\pi_{i+1} - \pi_i^*)' Q_1^{-1} G_i Q_1^{-1}(\pi_{i+1} - \pi_i^*). \qquad \text{(c)}
\end{aligned}$$

In view of Eqs. (b) and (c), Eq. (a) can be written as

$$\begin{aligned}
\tilde{J}_i = {} & [(\pi_i - \pi_i^*)' - G_i Q_1^{-1}(\pi_{i+1} - \pi_i^*)]' \\
& \times G_i^{-1}[(\pi_i - \pi_i^*)' - G_i Q_1^{-1}(\pi_{i+1} - \pi_i^*)] \\
& + (\pi_{i+1} - \pi_i^*)'(Q_1^{-1} + H_{i+1}' Q_2^{-1} H_{i+1} - Q_1^{-1} G_i Q_1^{-1})(\pi_{i+1} - \pi_i^*) \\
& + (x_{i+1} - H_{i+1}\pi_i^*)' Q_2^{-1}(x_{i+1} - H_{i+1}\pi_i^*) \\
& - 2(\pi_{i+1} - \pi_i^*)' H_{i+1}' Q_2^{-1}(x_{i+1} - H_{i+1}\pi_i^*).
\end{aligned}$$

Integration with respect to π_i yields constant $\int \exp\left(-\frac{1}{2}\tilde{J}_i\right) \mathrm{d}\pi_i$ = constant $\exp\left(-\frac{1}{3}\dddot{J}_i\right)$ where

$$\begin{aligned}
\dddot{J}_i = {} & \left(\pi_{i+1} - \pi_i^*\right)'(Q_1^{-1} - Q_1^{-1} G_i Q_1^{-1} + H_{i+1}' Q_2^{-1} H_{i+1}) \left(\pi_{i+1} - \pi_i^*\right) \\
& + \left(x_{i+1} - H_{i+1}\pi_i^*\right)' Q_2^{-1} \left(x_{i+1} - H_{i+1}\pi_i^*\right) \\
& - 2\left(\pi_{i+1} - \pi_i^*\right)' H_{i+1}' Q_2^{-1} \left(x_{i+1} - H_{i+1}\pi_i^*\right).
\end{aligned}$$

Hence

$$p(\pi_{i+1}|x^{i+1}) = \text{constant}\exp\left(-\frac{1}{2}\dddot{J}_i\right).$$

Since $p(\pi_{i+1}|x^{i+1})$ is proportional to the likelihood function, by maximizing the conditional density function, we are also maximizing the likelihood in order to determine π^*_{i+1}. Note that minimization of $\dddot{J}_i$, is equivalent to maximizing $p(\pi_{i+1}|x^{i+1})$. To minimize $\dddot{J}_i$, we expand Eq. (c) eliminating terms not containing π_{i+1}. Then, we differentiate with respect to π'_{i+1} and after equating to zero, we finally obtain (Lazaridis, 1980).

$$\pi^*_{i+1} = \pi^*_i + (Q_1^{-1} - Q_1^{-1}G_iQ_1^{-1} + H'_{i+1}Q_2^{-1}H_{i+1})^{-1} \times H'_{i+1}Q_2^{-1}(x_{i+1} - H_{i+1}\pi^*_i). \qquad \text{(d)}$$

Now, consider the composite matrix:

$$Q_1^{-1} - Q_1^{-1}G_iQ_1^{-1} \quad \text{where } G_i = (S_i^{-1} + Q_1^{-1})^{-1}.$$

Considering the matrix identity of householder, we can write

$$Q_1^{-1} - Q_1^{-1}(S_i^{-1} + Q_1^{-1})^{-1}Q_1^{-1} = (Q_1 + S_i)^{-1} \triangleq P_{i+1}^{-1}.$$

Hence Eq. (d) takes the form:

$$\pi^*_{i+1} = \pi^*_i + K_{i+1}(x_{i+1} - H_{i+1}\pi^*_i)$$

where K_{i+1}, S_{i+1}^{-1} and P_{i+1}^{-1} are defined in Eqs. (21)–(23).

It is recalled that H_0 is a null matrix, since no observations exist beyond period 1 in the estimation process. The same applies for the vector x_0.

Appendix B

It is recalled that the model was estimated using FIML method. The FIML estimates are as follows (R^2 and adjusted R^2 refer to the corresponding 2 *SLS* estimates. Numbers in brackets are the corresponding t-statistics).

Estimated Model

1. $$A_t = \underset{(3.48)}{0.842}Y_t + \underset{(1.74)}{0.024}A_{t-1} - \underset{(1.34)}{1.319}IR_t + \underset{(1.16)}{1.051}IR_{t-t} - \underset{(1.29)}{284}EXR_t + \underset{(4.87)}{55191.5}$$

$$R^2 = 0.99, \quad \overline{R}^2 = 0.92, \quad DW = 1.76, \quad \rho = 0.27$$

2. $(Y_t) = A_t + R_t$
3. $(BD)_t = (G_t + LR_t + PF_t) - (TY_t + GBS_t + AF_t + FB_t)$
4. $PF_t = \underset{(1.48)}{1148.80} + \underset{(2.39)}{0.169\,CFB_{t-1}} + \underset{(1.89)}{144.374\,WIR_t}$
5. $GBS_t = \underset{(1.14)}{0.641G_t} + \underset{(1.41)}{0.677G_{t-1}} + \underset{(1.64)}{1.591IR_t} - \underset{(0.23)}{0.33AF_t}$

$$R^2 = 0.89, \quad \overline{R}^2 = 0.84, \quad DW = 2.36, \quad \rho = \underset{(1.51)}{0.23}$$

6. $\Delta MD = \Delta MS$
7. $(MS)_t = [(1 + CD_t)/(CD_t + RR_t)](\Delta R_t + NDA_t)$
8. $RR_t = \underset{(3.47)}{0.008Y_t} - \underset{(-1.84)}{0.002Y_{t-1}} + \underset{(1.87)}{1.2571R_t} + \underset{(1.88)}{2.7031R_{t-1}} - \underset{(2.87)}{0.003T}$ $+ \underset{(2.72)}{0.065}$

$$R^2 = 0.86, \quad R^2 = 0.79, \quad DW = 2.88, \quad \rho = \underset{(1.23)}{0.26}$$

9. $CD_t = \underset{(-1.49)}{0.4391IR_t} + \underset{(1.16)}{0.158CD_{t-1}} + \underset{(1.56)}{0.0007Y_t} + \underset{(1.73)}{0.009Y_{t-1}}$ $- \underset{(-6.269)}{0.0057T} + \underset{(11.95)}{0.193}$

$$R^2 = 0.94, \quad \overline{R}^2 = 0.92, \quad DW = 1.22, \quad \rho = \underset{(4.21)}{0.64}$$

10. $MD_t = \underset{(-2.52)}{2.733RR_t} - \underset{(-0.24)}{2.19IR_t} + \underset{(2.17)}{1.713IR_{t-1}} + \underset{(6.67)}{1.275Y_t} - \underset{(-0.088)}{113335.0}$

$$R^2 = 0.86, \quad \overline{R}^2 = 0.81, \quad DW = 1.92, \quad \rho = \underset{(1.31)}{0.26}$$

11. $IR_t = \underset{(1.93)}{0.413MD_{t-1}} - \underset{(1.403)}{0.814IR_{t-1}} + \underset{(1.68)}{8.656CI_t} - 1.351CI_{t-1}$ $+ \underset{(1.56)}{0.406Y_{t-1}} - \underset{(-1.84)}{7.02}$

$$R^2 = 0.98, \overline{R}^2 = 0.95, DW = 2.48, \rho = \underset{(3.21)}{0.58}$$

12. $P_t = \underset{(-5.71)}{0.0004A_t} + \underset{(1.77)}{0.0002A_{t-1}} + \underset{(1.48)}{0.105IMC_t} + \underset{(3.79)}{0.421P_{t-1}} + \underset{(1.87)}{32.895}$

$$R^2 = 0.98, \overline{R}^2 = 0.93, DW = 2.09, \rho = 0.48$$

13. $IMC_t = \underset{(1.34)}{5.347} + \underset{(2.03)}{19.352WPM_t} + \underset{(0.97)}{9.017EXR_t}$

$$R^2 = 0.98, \quad \overline{R}^2 = 0.97, \quad DW = 2.07, \quad \rho = \underset{(1.37)}{0.31}$$

14. $R_t = X_t - IM_t + K_t + PFT_t + FB_t - PF_t + AF_t$

15. $$IM_t = \underset{(-0.82)}{-24.04} - \underset{(-0.86)}{0.025IM_{t-1}} + \underset{(1.59)}{0.104Y_t} - \underset{(9.21)}{0.089Y_{t-1}} - \underset{(-1.26)}{0.654IMC_t} + \underset{(0.27)}{1.123T}$$

$$R^2 = 0.96, \quad \overline{R}2 = 0.92, \quad DW = 2.52, \quad \rho = \underset{(-1.72)}{-0.62}$$

16. $CFB_t = \sum_{r=-20}^{t} FB_r$

$\lambda^2 = 563.09$

The estimated Chi-square satisfies the goodness of fit test for the FIML estimation.

Notations

A = Domestic absorption
AF = Foreign receipts (grants etc.)
BD = Government budget deficits
CD = Credit to deposit ratio in the commercial banking sector
CFB = Cumulative foreign borrowing, i.e., foreign debt over a period of 20 years
CT = Discount rate of the RBI
FB = Foreign borrowing
G = Government expenditure
GBS = Government bond sales
IM = Value of imports
IMC = Import price index (1990 = 100)
IR = Interest rate in the money-market
K = Foreign capital inflows
LR = Lending (minus repayments to the states)
MD = Money demand
MS = Money supply
NDA = Net domestic asset creation by the RBI
P = Consumers' price index (1990 = 100)
PFT = Private foreign transactions
PF = Foreign payments
R = Changes in foreign exchange reserve
RR = Reserve to deposit ratio in the commercial banking sector
TY = Government tax revenue
T = Time trend

WPM = World price index of India's imports (1990 = 100)
WIR = World interest rate, average of European and US money market rate
EXR = Exchange rate (Rs/US$)
X = Value of exports
Y = GNP at constant 1990 prices

Chapter 3

Mathematical Modeling in Macroeconomics

Topic 1

The Advantages of Fiscal Leadership in an Economy with Independent Monetary Policies

Andrew Hughes Hallet
James Mason University, USA

1. Introduction

In January 2006, Gordon Brown (as Britain's finance minister) was widely criticized by the European Commission and other policy makers for running fiscal policies, which they considered to be too loose and irresponsible. The UK fiscal deficit had breached the 3% of GDP that is considered to be safe. To make its point, the European Commission initiated an "excessive deficit" procedure against the UK government, claiming that its fiscal policies constituted a danger for the good performance of the UK economy and its neighbors, if these deficits were not contained and reversed. Yet the United Kingdom, alone among its European partners, allows fiscal policy to lead in order to achieve certain social expenditure and medium-term output objectives and a degree of co-ordination with the monetary policies designed to reach a certain inflation target. And the economic performance has been no worse, and may have been better than elsewhere in the Eurozone. Was Gordon Brown's strategy so mistaken after all?

The British fiscal policy has changed radically since the days when it tried to micro-manage all of the aggregate demand with an accommodating monetary policy in the 1960s and 1970s; and again during the 1980s, it was designed to strengthen the economy's supply-side responses, while the monetary policies were intended to secure lower inflation.

The 1990s saw a return to more activist fiscal policies — but policies designed strictly in combination with an equally active monetary policy based on inflation targeting and an independent Bank of England. They are set, in the main, to gain a series of medium- to long-term objectives — low debt, the provision of public services and investment, social equality, and economic efficiency. The income stabilizing aspects of the fiscal policy have, therefore, been left to act passively through the automatic stabilizers, which are part of any fiscal system, while the discretionary part (the bulk of the policy measures) is set to achieve these long-term objectives. Monetary policy, meanwhile, is left to take care of any short-run stabilization around the cycle; that is, beyond what, predictably, would be done by the automatic stabilizers.[1]

To draw a sharp distinction between actively managed long-run policies, and non-discretionary short-run stabilization efforts restricted to the automatic stabilizers, is of course the strategic policy prescription of Taylor (2000). Marrying this with an activist, monetary policy directed at cyclical stabilization, but based on an independent Bank of England and a monetary policy committee with the instrument (but not target) independence, appears to have been the innovation in the UK policies. It implies a leadership role for the fiscal policy, which allows both fiscal and monetary policies to be better co-ordinated — but without either losing their ability to act independently.[2]

Thus, Britain appears to have adopted a Stackelberg solution, which lies somewhere between the discretionary (but Pareto superior) co-operative solution, and the inferior but independent (non-co-operative) solution, as shown in Fig. 1.[3] Nonetheless, by forcing the focus onto long-run objectives, to the exclusion of the short term, this set-up has imposed a degree of pre-commitment (and potential for electoral punishment) on fiscal policy because governments naturally wish to lead. But, the regime remains non-co-operative so that there is no incentive to renege on earlier plans in the

---

[1]The Treasury estimates that the automatic stabilizers will, in normal circumstances, stabilize some 30% of the cycle; the remaining 70% being left to monetary policy (HM Treasury, 2003). The option to undertake discretionary stabilizing interventions is retained "for exceptional circumstances" however. Nevertheless, the need for any such additional interventions is unlikely: first, because of the effectiveness of the forward looking, symmetric and an activist inflation targeting mechanism adopted by the Bank of England; and, second, because the long-term expenditure (and tax) plans are deliberately constructed in nominal terms so that they add to the stabilizing power of the automatic stabilizers in more serious booms or slumps.

[2]For details on how this leadership vs. stabilization assignment is intended to work, see HM Treasury (2003) and Section 3 below. Australia and Sweden operate rather similar regimes.

[3]Figure 1 is the standard representation of the outcomes of the different game theory equilibria when the reaction functions form an acute angle. The latter is assured for our context when both sets of policy makers have some interest in inflation control and output stabilization, and the policy multipliers have the conventional signs (Hughes Hallett and Viegi, 2002). Stackelberg solutions, therefore, imply a degree of co-ordination in the outcomes, relative to the non-co-operative (unco-ordinated) Nash solution.

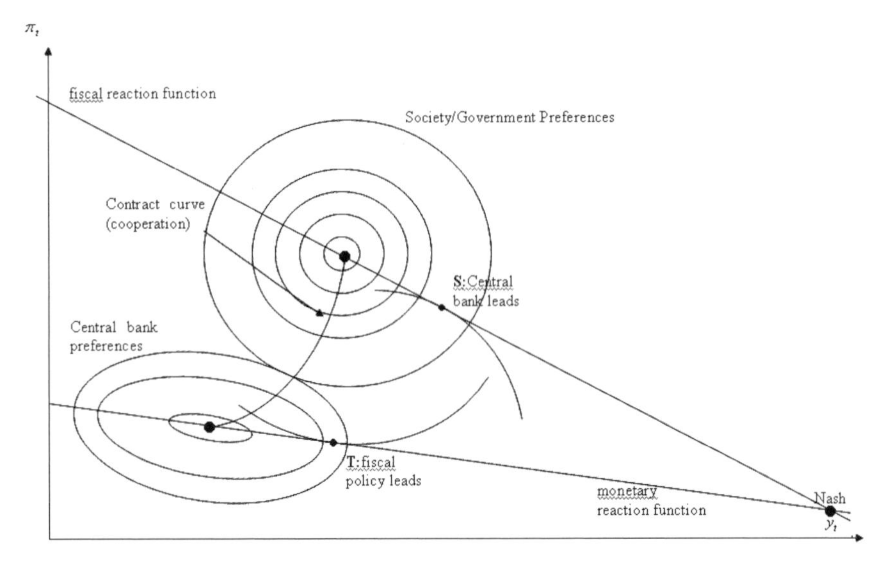

Figure 1. Monetary-Fiscal Interactions and the Central Bank.

absence of changes in information. Thus, the policies and their intended outcomes will be sustained by the government of the day.[4]

2. The Fiscal Leadership Hypothesis

The hypothesis is that the United Kingdom's improved performance is due to the fact that the fiscal policy leads an independent monetary policy. This leadership derives from the fact that fiscal policies typically have long-run targets (sustainability and low debt), and is not easily reversible (public services and social equality), and does not stabilize well if efficiency is to be maintained. Nevertheless, there are also automatic stabilizers in any fiscal policy framework, implying that monetary policy must condition itself on the fiscal stance at each point. This automatically puts the latter in a follower's role. This is helpful because it allows the economy to secure the benefits of an independent monetary policy *but also* enjoys a certain measure of co-ordination between the two sets of policy makers — discretionary/automatic fiscal policies on one side, and active monetary policies on the

[4]Stackelberg games, with fiscal policy leading, produce fiscal commitment: i.e., sub-game perfection with either strong or weak time consistency (Basar, 1989). In this sense, we have "rules rather than discretion". Commitment to the stabilization policies is then assured by the independence of the monetary authority.

other. The extra co-ordination arises in this case because the constraints on, and responses to an agreed leadership role reduce the externalties of self-interested behavior, which independent agents would otherwise impose on one another. This allows a Pareto improvement over the conventional non-co-operative (full independence) solution, without reducing the central bank's ability to act independently. It is important to realize that the co-ordination here is implicit and therefore rule-based, not discretionary.

To show that the UK fiscal policy does lead monetary policy in this sense, and that this is the reason for the Pareto improved results in Table 1, I produce evidence in three parts:

- *Institutional evidence*: taken from the UK Treasury's own account of how its decision-making works, what goals it needs to achieve, and how the monetary stance may affect it or vice versa;

Table 1. Generalized Taylor rules in the UK and EU-12.

S.No.	Const	r_{t-1}	$\pi^e_{j,k}$	j, k	gap	pd	debt
Dependent variable: central bank lending rate, r_t:							
For the UK, monthly data from June 1997–January 2004							
(1)	−1.72	0.711	1.394	+6, +18	0.540	—	—
	(2.16)	(5.88)	(2.66)		(2.06)		
	$\overline{R}^2 = 0.91, \quad F_{3,21} = 82.47, \quad N = 29$						
(2)	−2.57	0.598	1.289	+9, +21	1.10	−0.67	0.43
	(2.59)	(4.38)	(0.66)		(1.53)	(0.83)	(0.50)
	$\overline{R}^2 = 0.90, \quad F_{5,19} = 44.1, \quad N = 25$						
For the Eurozone (EU-12), monthly data from January 1999–January 2004							
(1)	−0.996	0.274	1.714	+9, +21	0.610	—	—
	(0.76)	(1.04)	(3.89)		(1.77)		
	$\overline{R}^2 = 0.82, \quad F_{3,11} = 22.2, \quad N = 15$						
(2)	−13.67	0.274	1.110	−6, +6	0.341	0.463	0.191
	(0.59)	(1.04)	(1.19)		(1.22)	(2.89)	(0.63)
	$\overline{R}^2 = 0.87, \quad F_{5,13} = 24.7, \quad N = 19$						

Notation: j, k give the bank's inflation forecast interval; $\pi^e_{j,k}$ represents the average inflation rate expected over the interval $t + j$ to $t + k$: $E\pi_{t+j,t+k}$; gap = GDP–trend GDP; pd = primary deficit/surplus as a percentage of GDP (a surplus > 0) and debt = debt/GDP ratio. Estimation: instrumented 2SLS; t-ratios in parentheses; j, k determined by search; and the output gap is obtained from a conventional HP filter to determine trend output.

- *Empirical evidence*: the extent to which the monetary policy is affected by the fiscal policy, but fiscal policy with its longer objectives does not depend on monetary policies and
- *Theoretical evidence*: shows how, with different goals for governments and central banks, Pareto improving results can be expected from fiscal leadership. The United Kingdom has, therefore, had the incentive, and the capacity, to operate in this way.

3. Institutional Evidence: The Treasury's Mandate

3.1. *History*

The British fiscal policy has changed a great deal over the past 30 years. Fiscal policy was the principal instrument of the economic policy in the 1960s and 1970s, and was focused almost exclusively on demand management. Monetary policy and the provision of public services (subject to a lower bound) were essentially accommodating factors, since the main constraint was perceived to be the financing of persistent current-account trade deficits. The result of this was a short-term view, in which the conflicts between the desire for growth (employment) and the recurring evidence of overheating (trade deficits) led to an unavoidable sequence of "stop-go" policies. There were many (Dow, 1964), who argued that the real problem was a lack of long-run goals for the fiscal policy; and that the lags inherent in recognizing the need for a policy change and in implementing it through Parliament till it takes hold in the markets, had produced a system that was actually destabilizing rather than stabilizing. But, there would have been conflicts anyway because there were two or more competing goals, but effectively only one instrument to reach them.

Such a system could not provide the long-run goal of public services, public investment, and social equity. It certainly proved too difficult to turn the fiscal policy on and off as fast and frequently as required, *and* pre-commit to certain expenditure and taxation plans at the same time. The point here is that, if you cannot pre-commit fiscal policy to certain goals, then you will not be able to pre-commit monetary policy either. This means that the "stable growth with low inflation" objective will be lost.

In the 1980s, the strategy changed when a new regime took office, committed to reducing the size and role of the government in the economy. The demand management role of the fiscal policy was phased out. This role passed over to inflation control and monetary management — with limited success in the earlier phases of monetary and exchange rate targeting, but with rather more success when it was formalized as an inflation-targeting regime under an independent Bank of England and monetary policy committee. In this period, therefore, the role of fiscal policy was to provide the "conditioning information" within which the monetary management had to work. It was split into two distinct parts. Short-term stabilization of output and employment was left to the automatic stabilizers inherent in the prevailing tax and expenditures rules. Discretionary adjustments would only be used in exceptional circumstances, if at all.[5] The rest of the fiscal policy, being the larger part, could then be directed at long-term objectives: in this instance, changes to free up the supply side (and enhance competitiveness) on the premise that greater microeconomic flexibility would both reduce the need for stabilizing interventions (relative wage and price adjustments would do the job better) and provide the conditions for stronger employment and growth (HMT, 2003). Given this, the long-run goals of better public services, efficiency, and social equity could then be attained. In addition, as the market flexibility changes were made, this part of fiscal policy could be increasingly focused on providing the long-run goals directly.[6]

3.2. *Current Practice*

According to the treasury's own assessment (HMT, 2003), the UK fiscal policy now leads monetary policy in two senses. First, fiscal policy is decided in advance of monetary or other policies (pp. 5, 63, 64, 74),[7] *and* with a long-time horizon (pp. 5, 42, 48, 61). Second, monetary policy is charged with controlling inflation and stabilizing output around the cycle — rather than steering the economy as such (pp. 61–63). Fiscal policy, therefore, sets the conditions within which monetary policy has to respond and achieve its own objectives (pp. 9, 15, 67–68). The short-term fiscal interventions, now less than half the total and declining (p. 48, Box 5.3, Section 6), are restricted to the automatic stabilizers — with effects that are known and predictable.

[5]This follows the recommendations in Taylor (2000). However, because of the inevitable trade-off between cyclical stability and budget stability, it is likely to be successful only in markets with sufficiently flexible wages and prices (see Fatas *et al.*, 2003).

[6]This summary is taken from the UK Treasury's own view (HMT, 2003), Sections 2, 4, and pp. 34–38.

[7]In this section, page or section numbers given without further reference are all cited from HMT (2003).

The short-run discretionary components are negligible (p. 59, Table 5.5), and the long-run objectives of the policy will always take precedence in cases of conflict (pp. 61, 63–68). Fiscal policy would, therefore, not be used for fine tuning (pp. 11, 63), or for stabilization (pp. 1, 14). This burden will be carried by interest rates, *given* the fiscal stance and its forward plans (pp. 1, 7, 11, 37).

Third, given the evidence that consumers and firms often do not look forward efficiently and may be credit-constrained, and also that the impacts of the fiscal policy are uncertain and have variable lags (pp. 19, 26, 48; Taylor, 2000), it makes sense for the fiscal policy to be used consistently and sparingly and in a way that is clearly identified with the long-term objectives. The United Kingdom has determined these objectives to be (pp. 11–13, 39–41, 61–63, 81–82):

(a) The achievement of sustainable public finances in the long term; low debt (40% of GDP) and symmetric stabilization over the cycle.
(b) A sustainable and improving delivery of public services (health and education); improving competitiveness/supply-side efficiency in the long term; and the preservation of social (and inter-generational) equity and alleviation of poverty.
(c) Recognition that the achievement of these objectives is often contractual and cannot easily be reversed once committed. The long lead times needed to build up these programs mean that the necessary commitments must be made well in advance. Frequent changes would conflict with the government's medium-term sustainability objectives, and cannot be fulfilled. Similarly, changes in direct taxes will affect both equity and efficiency, and should seldom be made.
(d) The formulation of clear numerical objectives is consistent with these goals; a transparent set of institutional rules to achieve them; and a commitment to a clear separation of these goals from economic stabilization around the cycle.[8]
(e) To ensure that fiscal policy can operate along these lines, public expenditures are planned with fixed three-year departmental expenditure limits which, when combined with decision and implementation lags of up to two years, means that the bulk of the fiscal policy has to be planned

[8]The credibility of fiscal policy, and by extension an independent monetary policy, is seen as depending on these steps, *and* on *symmetric* stabilization. See also Dixit (2001), and Dixit and Lambertini (2003).

with a horizon of up to five years — vs a maximum of two years in the inflation targeting rules operated by the Bank of England. Moreover, these spending limits are constraints defined in nominal terms, so that they will be met. In addition, the Treasury operates a tax smoothing approach — having rejected "formulaic rules" that might have adjusted taxes or expenditures, or the various tax regulators, or credit taxes that could have stabilized the economy in the short term. The bulk of fiscal policy will remain focused on the medium to long term therefore.

To summarize, the institutional structure itself has introduced fiscal leadership. And, since the discretionary elements are smaller than elsewhere — smaller than the automatic stabilizers (Tables 1–5) and declining (Box 5.3) — something else (monetary policy) must have taken up the burden of short-run stabilization. This arrangement is, therefore, best modeled as a Stackelberg game, with institutional parameters for goals, independence, priorities etc. I shall argue that this has had the very desirable result of increasing the degree of co-ordination between policy makers without compromising their formal or institutional independence — or indeed, their credibility for delivering stability and low inflation. Constrained independence, in other words, designed to reduce the externalities, which each party might, otherwise, impose on the other.

4. Empirical Evidence

The next step is to establish whether the UK authorities have followed the leadership model described above. The fact that they say that they follow such a strategy does not prove that they do so. The difficulty here is that, although the asymmetry in *ex-ante* (anticipated) responses between the follower and leader is clear enough in the theoretical model — the follower expects no further changes from the leader after the follower chooses his reaction function, whereas the leader takes this reaction function into account — such an asymmetry and zero restriction will not appear in the ex-post (observed) responses that emerge in the final solution (Basar and Olsder, 1999; Brandsma and Hughes Hallett, 1984; Hughes Hallett, 1984). Since I have no data on anticipations, this makes it difficult to test for leadership directly. But, we can use indirect tests based on the degree of competition, or complimentarity, between the instruments.

4.1. *Monetary Responses*

For monetary policy, it is widely argued that the authorities' decisions can best be modeled by a Taylor rule[9]:

$$r_t = \rho r_{t-1} + \alpha E_t \pi_{t+k} + \beta gap_{t+h} \quad \alpha, \beta, \rho \geq 0 \tag{1}$$

where k, h represent the authorities' forecast horizon[10] and may be positive or negative. Normally, $\alpha > 1$ will be required to avoid indeterminancy: that is, arbitrary variations in output or inflation, as a result of unanchored expectations in the private sector. The relative size of α and β then reveals the strength of the authorities' attempts to control inflation vs. income stabilization; and ρ their preference for gradualism. I set $h = 0$ in Eq. (1), since monetary policy appears not to depend on the expected output gaps (Dieppe *et al.*, 2004).

In order to obtain an idea of the influence of fiscal policies on monetary policy, I include some Taylor rule estimates — with and without fiscal variables — in Table 1. They show such rules for the United Kingdom and the Eurozone since 1997 and 1999, respectively, the dates when new policy regimes were introduced. The Eurozone has been included to emphasize the potential contrast between fiscal leadership in the United Kingdom, and the lack of it in Europe.

4.2. *Different Types of Leadership*

Conventional wisdom would suggest that Europe has either monetary leadership or independent policies, and hence policies which are *either* jointly dependent in the usual way, *or* which are complementary and mutually supporting. The latter implies that the monetary policy tends to expand/contract whenever fiscal policy needs to expand or contract — but not necessarily *vice versa*, when money is expanding or contracting and is sufficient to

[9]Taylor (1993b). One can argue that the policy should be based on fully optimal rules (Svensson, 2003), of which Eq. (1) will be a special case. But, it is hard to argue that policy makers actually do optimize when the additional gains from doing so may be small and when the uncertainties in their information, policy transmissions, or the economy's responses may be quite large. In practice, therefore, the Taylor rule approach is often found to fit central bank behavior better.

[10]In principle, k and h may be positive or negative: positive if the policy rule is based on future-expected inflation, to head off an anticipated problem, as in the Bank of England. But negative if interest rates are to follow a feedback rule to correct past mistakes or failures.

control inflation on its own. This is a weak form of monetary leadership in which fiscal policies are an additional instrument for use in cases of particular difficulty, rather than the policies in a Nash game with conflicting aims that need to be reconciled.

More generally, leadership implies complimentarity among policy instruments in the leader's reaction function, but conflicts among them in the follower's responses. A weak form of leadership also allows for independence among instruments in the leader's policy rules. Thus, monetary leadership would imply some complimentarity (or independence) in the Taylor rule, but conflicts in the fiscal responses. And fiscal leadership would mean complimentarity or independence in the fiscal rule, but conflicts in the monetary responses. Evidently, from Section 3, we might expect Stackelberg leadership (with fiscal policy leading) in the UK, but the opposite in the Eurozone.

4.3. *Observed Behavior*

The upper equations in each panel of Table 1 yield the standard results for the monetary behavior in both the United Kingdom and the Eurozone. Both monetary authorities have targeted expected inflation more than the output gap since the late 1990s — and with horizons of 18–21 months ahead. The European Central Bank (ECB) has been more aggressive in this respect. But, contrary to conventional wisdom, it was also more sensitive to the output gap and had a longer horizon and less policy inertia.

However, if we allow monetary policies to react to the changes in fiscal stance, we get different results (the lower equations). Here, we see that the UK monetary decisions may take fiscal policy into account, but the effect is *not* significant or well defined. However, this model of monetary behavior does imply more activist policies, a longer forecast horizon (up to two years as the Bank of England claims) and greater attention to the output gap — the symmetry in the UK's policy rule. And to the extent that fiscal policy does have an influence, it would be as a substitute (or competitor) for monetary policy — fiscal deficits lead to higher interest rates. This is potentially consistent with fiscal leadership, since this form of leadership can allow independence *or* complimentarity between instruments in the leader's rule. We need to check the fiscal reaction functions directly. The lack of significance for the debt ratio is easily understood, however. Since this is a declared long-run objective of the fiscal policy, it would not be necessary for the monetary policy to take it into account. So far, the evidence could imply a Stackelberg follower role or independence for monetary policy, easing the competition (externalities) among policy instruments.

The ECB results look quite different. Once fiscal effects are included, the concentration on inflation control is much reduced and the forecast horizon shrinks to six months. Moreover, a feedback element, to correct the past mistakes, comes in. At the same time, output stabilization becomes less important, which implies that the symmetric targeting goes out. Instead, monetary policy now appears to react to fiscal policy, but with the "wrong" sign: the larger the primary deficit, the looser will be the monetary policy. In this case, therefore, the policies are acting as complements — circumstances, which call for a primary deficit will also call for a relaxation of the monetary policy. Thus, we have an evidence of monetary leadership, where fiscal policy is used as an additional policy instrument.

4.4. *Fiscal Reaction Functions in the UK and Eurozone*

Many analysts have hypothesized that fiscal policy responses can best be modeled by means of a "fiscal Taylor rule":

$$d_t = a_t + \gamma\ gap_t + sd \quad \gamma > 0 \tag{2}$$

where sd = structural deficit ratio, d_t is the actual deficit ratio ($d > 0$ denotes a surplus),[11] and a_t represents all the other factors such as the influence of monetary policy, existing or anticipated inflation, the debt burden, or discretionary fiscal interventions. The coefficient γ then gives a measure of an economy's automatic stabilizers. The European Commission (2002), for example, estimates $\gamma \approx 1/2$ for Europe — a little more in countries with an extensive social security system, a little less elsewhere. And a similar relationship is thought to underlie UK's fiscal policy (see HMT, 2003: Boxes 5.2 and 6.2).

The expected signs of any remaining factors are not so clear. The debt burden should increase current deficits, unless there is a systematic debt reduction program underway. Inflation should have a positive impact on the deficit ratio if fiscal policy is used for stabilization purposes, but it had no effect otherwise. Finally, the output gap should also have a positive impact on the deficit ratio, if the latter is being used for stabilization purposes — in which case interest rates should be negatively correlated with the size

[11] See Taylor (2000), Gali and Perotti (2003), Canzoneri (2004), and Turrini and in 't Veld (2004) for similar formulations. The European Commission (2002) uses a similar rule, but defines $d > 0$ to be a deficit. They therefore expect γ to be negative.

Table 2. Fiscal policy reaction functions.

For the United Kingdom, sample period 1997q3–2004q2

$$d_t = \underset{(0.37)}{-0.444} + \underset{(7.64)}{0.845 debt_t} + \underset{(0.30)}{0.0685 r_t} + \underset{(1.72)}{1.076 gap_t} \qquad \overline{R}^2 = 0.78, \quad F_{3,24} = 33.23$$

For the Eurozone, sample period 1999q1–2004q2

$$d_t = \underset{(2.96)}{3.36} + \underset{(9.66)}{1.477 d_{t-1}} - \underset{(2.21)}{0.740 \pi^e_{+9,21}} - \underset{(1.16)}{0.207 r_t} - \underset{(2.18)}{0.337 gap_t} \qquad \overline{R}^2 = 0.95, \quad F_{4,11} = 54.55$$

Key: d_t = gross deficit/GDP (%), where $d < 0$ denotes a deficit; $debt_t$ = debt/GDP (%); r_t is the central bank lending rate; $\pi^e_{j,k}$ and gap_t as in Table 2(a).
Estimation method, instrumented 2SLS; linear interpolation for quarterly deficit figures; and t-ratios in parentheses.
Note: At current debt, interest rates and inflation targets, these estimates imply a structural deficit of about 0.1% for the United Kingdom and 2.6% for the Eurozone.

of the deficit, because monetary policies focused on inflation, and fiscal policies focused on short-run stabilization, would conflict. Conversely, a negative association with the output gap, but a positive one with interest rates, would imply no automatic stabilizer effects but mutually supporting policies: i.e., higher interest rates go with tighter fiscal policies. This implies complementary policies.

Table 2 includes our estimates of the fiscal policy reaction functions for the United Kingdom and the Eurozone. Higher debt increases the surplus ratio in the United Kingdom, but has no effect in the Euro area. So, while the United Kingdom, evidently, has had a systematic debt reduction program, no such efforts have been made in Europe. Inflation, on the other hand, has had no effect on the UK deficit, but a negative one in Europe. Similarly, the output gap produces a negative reaction in Europe, but a positive one in the United Kingdom. These two variables, therefore, indicate that the fiscal policy has been used for output stabilization in the United Kingdom — consistent with allowing automatic stabilizers to do the job — but for purposes other than this in the Eurozone. This result fits in neatly with Europe's evident inability to save for a rainy day in the upturn, *and* the inability to stabilize in the downturn because of the stability pact.[12]

[12] Buti *et al.* (2003). The presence of debt in the UK rule indicates that sustainability has been a primary target in the United Kingdom, but not in the Eurozone.

Consequently, the United Kingdom appears to use fiscal policy for stabilization in a minor way as claimed, while the Euro economies seem not to use fiscal policy this way at all. Confirmation of this comes from the responses to interest rate changes, which are positive but very small and statistically insignificant in the United Kingdom, but negative and near significant in Europe. This implies that the British fiscal policies are chosen independently of the monetary policy, as suggested by our weak leadership model. But, if there is any association at all, then both the fiscal and monetary policies would be compliments and weakly co-ordinated.

The UK results are therefore inconclusive. They are consistent with independent policies, or fiscal leadership — the latter, despite the insignificance of the direct fiscal-monetary linkage, because of the significant output gap term in the fiscal equation which, given the same effect is not found in the monetary policy reactions, suggests fiscal leadership may in fact have been operating. In any event, there is no suggestion of monetary leadership; if anything happens, the results imply independence or fiscal leadership.

In the Eurozone, the results are quite different. Here, the significant result is the conflict among instruments in monetary policy (and essentially the same conflict in the fiscal policy reactions). Hence, the policies are competitive, which suggests that they form a Nash equilibrium (or possibly monetary leadership, since the coefficient on fiscal policy in the Taylor rule is small and there is no output gap smoothing). This suggests weak monetary leadership, or a simple non-co-operative game.

5. Theoretical Evidence: A Model of Fiscal Leadership

5.1. *The Economic Model and Policy Constraints*

The key question now is: would governments want to pursue fiscal precommitment? Do they have an incentive to do so? And would there be a clear improvement in terms of an economic performance if they did? More important, would the fiscal leadership model be more advantageous, if fiscal policy was limited by a deficit rule in the form of "hard" targets (as in the original stability pact) or "soft" targets?

To answer these questions, we extend a model used in Hughes Hallett and Weymark (2002; 2004a;b; 2005) to examine the problem of monetary policy design when there are interactions with fiscal policy. For exposition purposes, we suppress the spillovers among countries and focus on the following three equations to represent the economic structure of any one country[13]:

$$\pi_t = \pi_t^e + \alpha y_t + u_t \quad (3)$$

$$y_t = \beta(m_t - \pi_t) + \gamma g_t + \varepsilon_t \quad (4)$$

$$g_t = m_t + s(by_t - \tau_t) \quad (5)$$

where π_t is inflation in period t, y_t is the growth in output (relative to trend) in period t, and π_t^e represents the rate of inflation that the rational agents expect to prevail in period t, conditional on the information available at the time expectations are formed. Next, m_t, g_t, and τ_t represent the growth in the money supply, government expenditures, and tax revenues in period t, and u_t and ε_t are random disturbances, which are distributed independently with zero mean and constant variance. All variables are deviations from their steady-state (or equilibrium) growth paths, and we treat trend budget variables as pre-committed and balanced. Deviations from the trend budget are, therefore, the only discretionary fiscal policy choices available. The coefficients α, β, γ, s, and b are all positive by assumption. The assumption that γ is positive is sometimes controversial.[14] However, the short-run impact multipliers derived from Taylor's (1993a) multicountry estimation provide an empirical support for this assumption (as does the HMT, 2003).

According to Eq. (3), inflation is increasing, as the rate of inflation predicted by private agents and in output growth. Equation (4) indicates that both monetary and fiscal policies have an impact on the output gap. The micro-foundations of the aggregate supply Eq. (3), originally derived by Lucas (1972; 1973), are well known. McCallum (1989) shows that aggregate demand equation like Eq. (4) can be derived from a standard, multi-period utility-maximization problem.

Equation (5) describes the government's budget constraint. In earlier chapters, we allowed *discretionary* tax revenues to be used for

[13]Technically, we assume a blockwise orthogonalization of the traditional multicountry model to produce independent semi-reduced forms for each country. The disturbance terms may, therefore, contain foreign variables, but they will have zero means so long as these countries remain on their long-run (equilibrium) growth paths on average (all variables being defined as deviations from their equilibrium growth paths).

[14]Barro (1981) argues that government purchases have a contractionary impact on output. Our model, by contrast, treats fiscal policy as important because: (i) fiscal policy is widely used to achieve re-distributive and public service objectives; (ii) governments cannot pre-commit monetary policy with any credibility if fiscal policy is not pre-committed (Dixit and Lambertini, 2003), and (iii) Central Banks, and the ECB, in particular, worry intensely about the impact of fiscal policy on inflation and financial stability (Dixit, 2001).

re-distributive purposes only, but permitted *discretionary* expenditures for enhancing output. We further assumed that there were two types of agents — rich and poor — and that only the rich use their savings to buy government bonds. Based on this view, b would be the proportion of pre-tax income (output) going to the rich and s, the proportion of after-tax income that the rich allocate to saving. Tax revenues, τ_t, can then be used by the government to re-distribute income from the rich to poor, either directly or via public services. This structure, therefore, has output-enhancing expenditures g_t, and discretionary transfers τ_t. Both are financed by aggregate tax revenues; that is, from discretionary and trend revenues. Expenditures above these revenues must be financed by the sale of bonds.

5.2. *An Alternative Interpretation*

We could, however, take a completely different interpretation of Eq. (5). We could take the term $s(by_t - \tau_t)$ to be the proportion of the budget-deficit ratio, as it currently exists, adjusted for the effect of growth on this ratio and any discretionary taxes raised in this period, which the government now proposes to spend in period t. The deficit ratio itself is of course $d = (G - T)/Y = (e - r)$, where G and T are the absolute levels of government spending and tax revenues, and e and r are their counterparts as a proportion of Y. If the economy grows, then $\Delta d = \Delta(G-T)/Y - \dot{y}(e-r)$ where $\dot{y}$ denotes the growth *rate* in national income. If we wish to see what would happen to this deficit ratio if fiscal policies were not changed, it means new spending can only be allowed to take place out of new revenues generated by growth: $\Delta G = \Delta T = r\Delta Y$. Inserting this, we have $\Delta d = -(e - r)\dot{y} = -d\dot{y}$. Since y_t is the deviation of Y from its steady-state path, by_t in Eq. (5) will equal Δd, the change in the existing deficit ratio under existing expenditure plans and tax codes, if $b = -(e - r)/Y_1$. The government will spend some proportion of that, s, less new discretionary taxes in the current period. Hence, s would typically equal 1, although it might be less, if some parts were saved in social security funds or other assets. This defines what would happen to the deficit ratio if there were no change in existing fiscal policies; but the term in τ_t shows that governments may also raise additional taxes in order to reduce the current deficit ratio to some desired target value, θ, say. This target is likely to be $\theta = 0$, to balance the budget over the cycle, as required by the stability pact. We give governments such a target, in the form of either a soft or a hard rule, in the objective functions that are discussed in Section 5.

Given this new interpretation, we can now solve for π_t^e, π_t, and y_t from Eqs. (3) and (4). This yields the following reduced forms:

$$\pi_t(g_t, m_t) = (1+\alpha\beta)^{-1}\left[\alpha\beta m_t + \alpha\gamma g_t + m_t^e + \frac{\gamma}{\beta}g_t^e + \alpha\varepsilon_t + u_t\right] \quad (6)$$

$$y_t(g_t, m_t) = (1+\alpha\beta)^{-1}[\beta m_t + \gamma g_t - \beta m_t^e - \gamma g_t^e + \varepsilon_t - \beta u_t]. \quad (7)$$

Solving for τ_t using Eqs. (5) and (7), then yields:

$$\tau_t(g_t, m_t) = [s(1+\alpha\beta)]^{-1}[(1+\alpha\beta+sb\beta)m_t - (1+\alpha\beta - sb\gamma)g_t - sb\beta m_t^e - sb\gamma g_t^e + sb(\varepsilon_t - \beta u_t)]. \quad (8)$$

5.3. *Government and Central Bank Objectives*

In our formulation, we allow for the possibility that the government and an independent central bank may differ in their objectives. In particular, we assume that the government cares about inflation stabilization, output growth, and the provision of public services (and hence the size of the public sector deficit or debt); whereas the central bank, if left to itself, would be concerned only with the first two objectives,[15] and possibly only the first one. We also assume that the government has been elected by majority vote, so that the government's loss function reflects society's preferences to a significant extent.

Formally, the government's loss function is given by:

$$L_t^g = \frac{1}{2}(\pi_t - \hat{\pi})^2 - \lambda_1^g y_t + \frac{\lambda_2^g}{2}[(b-\theta)y_t - \tau_t]^2 \quad (9)$$

where $\hat{\pi}$ is the government's inflation target, λ_1^g is the relative weight or importance that the government assigns to output growth,[16] and λ_2^g is the relative weight, which it assigns to the debt or deficit rule. The parameter θ represents the target value for the debt or deficit to the GDP ratio, which the government would like to reach: hence $(b-\theta)y_t$ becomes the target for its

[15]Since the central bank has no instruments to control the debt itself, it can only react to poor fiscal discipline indirectly: e.g., to the extent that its inflation objective is compromised, where it does have an instrument.

[16]Barro and Gordon (1983) also adopt a linear output target. In the delegation literature, the output component in the government's loss function is usually represented as quadratic to reflect an output *stability* objective. In our model, the quadratic term in debt/deficits allows monetary and fiscal policy to play a stabilization role as well as pick a position on the economy's output-inflation trade-off.

discretionary revenues, τ_t. All the other variables are as previously defined. We may regard large values of λ_2^g, say $\lambda_2^g > \max[1, \lambda_1^g]$, as defining a "hard" debt or deficit rule; and smaller values, $\lambda_2^g < \min[1, \lambda_1^g]$, as a "soft" debt or deficit rule.

The objectives of the central bank, however, may be quite different from those of the government. We model them as follows:

$$L_t^{cb} = \frac{1}{2}(\pi - \hat{\pi})^2 - (1-\delta)\lambda^{cb} y_t - \delta\lambda_1^g y_t + \frac{\delta\lambda_2^g}{2}[(b-\theta)y_t - \tau_t]^2 \tag{10}$$

where $0 \leq \delta \leq 1$, and λ^{cb} is the weight, which the central bank assigns to the output growth. The parameter δ measures the degree to which the central bank is forced to take the government's objectives into account. The closer δ is to 0, the greater is the independence of the central bank in making its choices. And the lower is λ^{cb} is, the greater is the degree of its conservatism in making these choices.

In Eq. (9), we have defined the government's inflation target as $\hat{\pi}$. The fact that the same inflation target appears in Eq. (10) would be correct for the cases where the bank has the instrument independence, but not target independence. However, it is easy to relax this assumption and allow the central bank to choose its own target, as the ECB does. But, as I show in Hughes Hallett (2005), there is no advantage in doing so since the government would simply adjust its parameters to compensate. Hence, only if the bank is free to choose the value of λ^{cb} as well, do we get an extra advantage. Yet, even this will not be enough to outweigh the advantages to be gained from fiscal leadership — for the reasons noted in Section 5.

A second feature is that Eqs. (9) and (10) specify symmetric inflation targets around $\hat{\pi}$. Symmetric inflation targets are particularly emphasized as being required of both monetary policy and the fiscal authorities (HMT, 2003). We have, therefore, specified both to be features of the government and the central bank objective functions here. However, it is not clear that symmetry has been a feature of the European policy making in practice. Indeed, the ECB has acquired a reputation for being more concerned to tighten monetary policy when $\pi > \hat{\pi}$, than it is to loosen it when $\pi < \hat{\pi}$. Consequently, rather than symmetry in the output-gap target, we have asymmetric penalties, which tighten the fiscal constraints in recessions, but weaken them in a boom when the debt and deficits would be falling anyway. On one hand, this might not be ideal, since it introduces an element of pro-cyclicality; on the other hand, it brings the rule closer to reality.

5.4. *Institutional Design and Policy Choices*

We characterize the strategic interaction between the government and the central bank as a two-stage non-co-operative game in which the structure of the model and the objective functions are common knowledge. In the first stage, the government chooses the institutional parameters δ and λ^{cb}. The second stage is a Stackelberg game in which fiscal policy takes on a leadership role. In this stage, the government and the monetary authority set their policy instruments, given the δ and λ^{cb} values determined at the previous stage. Private agents understand the game and form rational expectations for future prices in the second stage. Formally, the policy game runs as follows:

5.4.1. *Stage 1*

The government solves the problem:

$$\min_{\delta,\lambda^{cb}} EL^g(g_t, m_t, \delta, \lambda^{cb}) = E\left\{\frac{1}{2}[\pi_t(g_t, m_t) - \hat{\pi}]^2 - \lambda_1^g[y_t(g_t, m_t)]\right\} + \frac{\lambda_2^g}{2}E[(b-\theta)y_t(g_t, m_t) - \tau_t(g_t, m_t)]^2 \quad (11)$$

where $L^g(g_t, m_t, \delta, \lambda^{cb})$ is Eq. (9) evaluated at $(g_t, m_t, \delta, \lambda^{cb})$, and E denotes expectations.

5.4.2. *Stage 2*

(a) Private agents form rational expectations about future prices π_t^e, before the shocks u_t and ε_t are realized.
(b) The shocks u_t and ε_t are realized and observed by both the government and the central bank.
(c) The government chooses g_t, before m_t is chosen by the central bank, to minimize $L^g(g_t, m_t, \bar{\delta}, \bar{\lambda}^{cb})$ where $\bar{\delta}$ and $\bar{\lambda}^{cb}$ are at the values determined at stage 1.
(d) The central bank then chooses m_t, taking g_t as given, to minimize:

$$L^{cb}(g_t, m_t, \bar{\delta}, \bar{\lambda}^{cb}) = \frac{(1-\bar{\delta})}{2}[\pi_t(g_t, m_t) - \hat{\pi}]^2 - (1-\bar{\delta})\bar{\lambda}^{cb}[y_t(g_t, m_t)] + \bar{\delta}L^g(g_t, m_t, \bar{\delta}, \bar{\lambda}^{cb}) \quad (12)$$

We solve this game by solving the second stage (for the policy choices) first, and then substituting the results back into Eq. (11) to determine the optimal operating parameters δ and λ^{cb}. From stage 2, we get:

$$\pi_t(\delta, \lambda^{cb}) = \hat{\pi} + \frac{(1-\delta)\beta(\phi - \eta\Lambda)\lambda^{cb} + \delta(\beta\phi + \gamma\Lambda)\lambda_1^g}{\alpha[\beta(\phi - \eta\Lambda) + \delta\Lambda(\beta\eta + \gamma)]} \tag{13}$$

$$y_t(\delta, \lambda^{cb}) = -u_t/\alpha \tag{14}$$

$$\tau_t(\delta, \lambda^{cb}) = \frac{(1-\delta)\beta s(\beta\eta + \gamma)(\lambda^{cb} - \lambda_1^g)}{[\beta(\phi - \eta\Lambda) + \delta\Lambda(\beta\eta + \gamma)]\lambda_2^g} - \frac{(b-\theta)u_t}{\alpha} \tag{15}$$

where

$$\eta = \frac{\partial m_t}{\partial g_t} = \frac{-\alpha^2\gamma\beta s^2 + \delta\phi\Lambda\lambda_2^g}{(\alpha\beta s)^2 + \delta\Lambda^2\lambda_2^g} \tag{16}$$

$$\phi = 1 + \alpha\beta - \gamma\theta s \tag{17}$$

and

$$\Lambda = 1 + \alpha\beta + \beta\theta s. \tag{18}$$

Evidently, Λ is positive. We assume ϕ to be positive as well. One would certainly expect $\phi > 0$ since, with $\theta < 1$ and $s < 1$, fiscal policy would otherwise have to have such a strong impact on national income this, together with a Phillips curve that is sufficiently flat and weak monetary transmissions, government expenditures can simultaneously boost the output and be used to reduce deficit spending without worsening the budget (and subsequently debt) at the same time: $\gamma > (1 + \alpha\beta)/(\theta s)$. In practice, with $s \approx 1$, as noted above and a target deficit of zero, this would require very large fiscal multipliers indeed. In fact, numerical estimates for 10 larger OECD economies place ϕ very close to unity, rather than negative.[17] And with balanced budget targets ($\theta = 0$), we would certainly have $\phi > 0$. Nevertheless, the conflict which stands revealed within the fiscal policy is important. In order to get $\phi < 0$, output has to be capable of growing fast enough to generate sufficient revenues to boost output, when needed *and* to pay down the debt/deficit. If this is not possible, one will have to come at the expense of the other. This underlines the natural pro-cyclicality of *any* fiscal restraint mechanism under "normal" parametric values.[18]

[17] See Table 4 in Appendix A.

[18] Formally, $\partial\{(b-\theta)y_t - \tau_t]/\partial g_t = \phi/(1+\alpha\beta)$, which implies the gap between the fiscal adjustment needed and the pay-down allocated, will rise as more funds are used for stabilization, if $\phi > 0$.

Substituting Eqs. (13)–(15) back into Eq. (11), we can now get the stage 1 solution from:

$$\min_{\delta,\lambda^{cb}} EL^g(\delta,\lambda^{cb}) = \frac{1}{2}\left\{\frac{(1-\delta)\beta(\phi-\eta\Lambda)\lambda^{cb}+\delta(\beta\phi+\gamma\Lambda)\lambda_1^g}{\alpha[\beta(\phi-\eta\Lambda)+\delta\Lambda(\beta\eta+\gamma)]}\right\}^2 + \frac{\lambda_2^g}{2}\left\{\frac{(1-\delta)\beta s(\beta\eta+\gamma)(\lambda^{cb}-\lambda_1^g)}{[\beta(\phi-\eta\Lambda)+\delta\Lambda(\beta\eta+\gamma)]\lambda_2^g}\right\}^2. \quad (19)$$

This part of the problem has first-order conditions:

$$(1-\delta)(\phi-\eta\Lambda)\lambda_2^g\{(1-\delta)\beta(\phi-\eta\Lambda)\lambda^{cb}+\delta(\beta\phi+\gamma\Lambda)\lambda_1^g\} - (1-\delta)^2(\beta\eta+\gamma)^2\alpha^2 s^2\beta(\lambda_1^g-\lambda^{cb}) = 0 \quad (20)$$

and

$$\{(1-\delta)\beta(\phi-\eta\Lambda)\lambda^{cb}+\delta(\beta\phi+\gamma\Lambda)\lambda_1^g\} \times (\lambda_1^g-\lambda^{cb})\,\{\delta(1-\delta)\Lambda\Omega+(\phi-\eta\Lambda)\}\,\lambda_2^g - (1-\delta)(\beta\eta+\gamma)\alpha^2 s^2\beta\,\{(\beta\eta+\gamma)-(1-\delta)\beta\Omega\}\,(\lambda_1^g-\lambda^{cb})^2 = 0. \quad (21)$$

where $\Omega = \partial\eta/\partial\delta$. There are two real-valued solutions which satisfy this pair of first-order conditions.[19] Both are satisfied when $\delta = 1$ and $\lambda^{cb} = \lambda_1^g$. This solution describes a fully dependent central bank, which is not appropriate in the Eurozone case. And, it turns out to be inferior to the second solution: $\delta = \lambda^{cb} = 0$. In this solution, the central bank is fully independent and exclusively concerned with the economy's inflation performance.

Out of the two possibilities, the solution which yields the lowest welfare loss, as measured by the government's (society's) loss function, can be identified by comparing Eq. (19) to the expected loss that would be suffered under the alternative institutional arrangement. Substituting, $\delta = 1$ and $\lambda^{cb} = \lambda_1^g$ in Eq. (19) results in:

$$EL^g = \frac{(\lambda_1^g)^2}{2\alpha^2}. \quad (22)$$

[19] Because η is a function of δ, Eq. (21) is quartic in δ. This polynomial has four distinct roots, of which only two are real-valued. The complete solution may be found in Hughes Hallett and Weymark (2002).

But substituting, $\delta = \lambda^{cb} = 0$ in the right-hand-side of Eq. (19) yields:

$$EL^g = 0. \tag{23}$$

Consequently, our results show that, when there is fiscal leadership, society's welfare loss (as measured by Eq. (19)) is minimized when the government appoints independent central bankers who are concerned only with the achievement of a mandated inflation target and completely disregard the impact their policies may have on the output.

However, our results also indicate that fiscal leadership with an independent central bank can be beneficial under more general conditions. When $\delta = 0$, $\beta\eta + \gamma = 0$ and the externalities between policy makers are neutralized.[20] As a result, Eq. (19) will become:

$$EL^g = \frac{1}{2}\left\{\frac{\lambda^{cb}}{\alpha}\right\}^2 \tag{24}$$

for any value of λ^{cb}. Hence, an independent central bank will always produce better results than a dependent one, so long as it is more conservative than the government ($\lambda^{cb} < \lambda_1^g$), *irrespective* of the latter's commitment to the debt rule (λ_2^g). A conservative central bank will therefore be the best, but any bank that is more conservative than the government will do if the debt rule is to be effective.

A more interesting question is whether fiscal leadership with an independent central bank also produces better outcomes from a society's perspective, than those obtained in a simultaneous move game without leadership — the model generally favored in Europe. In the simultaneous move game, the solution to the stage 1 minimization problem is:

$$\delta = \frac{\beta\phi^2\lambda^{cb}\lambda_2^g + \alpha^2\gamma^2 s^2\beta(\lambda^{cb} - \lambda_1^g)}{\beta\phi^2\lambda^{cb}\lambda_2^g + \alpha^2\gamma^2\beta(\lambda^{cb} - \lambda_1^g) - \phi(\beta\phi + \gamma\Lambda)\lambda_1^g\lambda_2^g} \tag{25}$$

and a society's welfare loss will then be:

$$EL^g = \frac{1}{2}\left\{\frac{\lambda_1^g}{\alpha}\right\}\left\{\frac{(\alpha\gamma s)^2}{(\alpha\gamma s)^2 + \phi^2\lambda_2^g}\right\}. \qquad (26)^{21}$$

[20]Because $\beta\eta + \gamma = (\partial y/\partial m)\partial m/\partial g + \partial y/\partial g$, and the central bank has already taken the impact of its decisions on the government's decisions (also zero in this case) into account.

[21]Hughes Hallett and Weymark (2002; 2004a) derive these results in detail.

This is always smaller than the loss incurred when fiscal leadership is combined with a *dependent* central bank. However, the optimal degree of conservatism for an *independent* central bank, in this case, is obtained by setting $\delta = 0$ in Eq. (25) to yield:

$$\lambda^{cb*} = \frac{(\alpha\gamma s)^2 \lambda_1^g}{(\alpha\gamma s)^2 + \phi^2 \lambda_2^g}. \tag{27}$$

It is straightforward to show that the value of EL^g in Eq. (24) is always less than Eq. (26) as long as:

$$\lambda^{cb} < [\lambda_1^g \lambda^{cb*}]^{1/2}. \tag{28}$$

It is also evident that $\lambda^{cb*} < \lambda_1^g$ holds for *any* degree of commitment, however small, to the debt/deficit rule: $\lambda_2^g > 0$. Consequently, fiscal leadership with any $\lambda^{cb} < \lambda^{cb*}$ will always produce better outcomes, from a society's point of view, than any simultaneous move game between the central bank and the government. This is important because many inflation-targeting regimes, such as the ones operated by the Bank of England, the Swedish Riksbank, and the Reserve Bank of New Zealand, operate with fiscal leadership. By contrast, several others, notably the ECB and the US Federal Reserve System, do not operate with fiscal leadership. They are better thought of as being engaged in simultaneous move game with their governments.

5.5. *The Gains from Fiscal Leadership*

Finally, where do these leadership gains come from? Substituting $\delta = 0$ and $\lambda^{cb} = 0$ in Eqs. (13)–(15) yields:

$$\pi_t = \hat{\pi}, \quad y_t = -u_t/\alpha, \quad \text{and} \quad \tau_t = -(b - \theta)u_t/\alpha. \tag{29}$$

as final outcomes. By contrast, the outcomes for the simultaneous move policy game are:

$$\pi_t^* = \hat{\pi} + \frac{\alpha(\gamma s)^2}{[(\alpha\gamma s)^2 + \phi^2 \lambda_2^g]} = \hat{\pi} + \lambda^{cb*}/\alpha \tag{30}$$

$$y_t^* = -u_t/\alpha \tag{31}$$

$$\tau_t^* = \frac{\gamma s(\lambda^{cb*} - \lambda_1^g)}{\phi \lambda_2^g} - \frac{(b - \theta)u_t}{\alpha}. \tag{32}$$

Comparing these two sets of outcomes, we have five conclusions. These conclusions signify the main results of this paper:

(a) Fiscal leadership eliminates the inflationary bias, and results in lower inflation without any loss in output or output volatility. The proximate cause of this surprising result is that optimization under fiscal leadership leads to higher taxes and larger debt repayments.[22]

(b) The deeper reason is that there is a self-limiting aspect to the long-run design of fiscal policy. Unless the effect of fiscal policy on output is so large as to generate savings/taxes that could finance both fiscal expansions *and* debt re-payments (this possibility is ruled out in the deficit rule case), each expansion designed to increase output would simply be accompanied by a greater burden of debt. Hence, in order to preserve the long-run sustainability of its finances, the government must eventually raise taxes. This fact takes some inflationary pressure off the central bank when fiscal expansions are called for. As a result, the bank is less likely to tighten the monetary policy, and externalities are reduced on both sides. This is what generates a better co-ordination between the policy makers, as noted in Section 5. But, this can only happen if the government has a genuine commitment to the long-run goals (sustainable public finances, and a debt or deficit rule in this case), since only then will the fiscal policy leader take into account the predictable reactions (of the follower) to any short-run deviations by the leader. The follower knows that the leader is not likely to sacrifice his long-run goals by making such deviations (and anyway has the opportunity to clear up the mess, according to the follower's own preferences, if the leader does so). Thus, in the short-run under simultaneous moves, each player fails to consider the predictable response, and costs, which the externalities he/she imposes on his/her rival would create. The ability to co-ordinate would be lost, and the outcomes worse for both. We show this in Section 5 (and in a more formal analysis in Hughes Hallett, 2005).

(c) These results hold independently of the commitment to the debt rule (λ_2^g), or its target value (θ), and independently of the government's preference to stabilize or spend (λ_1^g, s), and of the economy's transmission parameters (α, β, γ) or the particular value of b as discussed above.

[22] Taxes are lower under simultaneous moves because $\lambda^{cb} < \lambda_1^g$. So $E\tau = 0$ in Eq. (29) vs. $E\tau_t^* \leq 0$ in Eq. (32).

(d) Since $E\tau_t^* < 0$ in Eq. (32), but $E\tau_t = 0$ in (29), the simultaneous moves regime will always end up in increasing the deficit levels. Therefore, it will eventually exceed any limit set for the debt ratio. Fiscal leadership will do neither of these things. More generally, the simultaneous moves game always leads to fiscal expansions if money financing is small.[23] Indeed, the expected value of $s(by_t^* - \tau_t^*)$ in Eq. (5) is zero under fiscal leadership, but $\phi\lambda^{cb*}/(\alpha^2\gamma) > 0$ in the simultaneous moves case (as expected, since the monetary policy externalities would be larger). And since revenues are lower in the latter case, budget deficits are larger. Hence, the gains from the fiscal leadership are more than just better outcomes. Leadership also implies less expansionary budgets and tighter public expenditure controls. The ECB should find this a significantly more comfortable environment to operate in.

(e) The simultaneous moves regime approaches fiscal leadership, as the debt rule becomes progressively "harder": that is, $EL^g \to 0$, $\lambda^{cb*} \to 0$; $\pi_t^* \to \hat{\pi}$, and $E\tau_t^* \to 0$, as $\lambda_2^g \to \infty$.

6. The Co-Ordination Effect

In our model, the central bank is independent. Without further institutional restraints, interactions between an independent central bank and the government would lead to non-co-operative outcomes. But, if the government is committed to long-term leadership in the manner that I have described, the policy game will become an example of *rule-based* co-ordination in which both parties can gain without any reduction in the central bank's independence. However, this is not to say that both parties gain equally. If the inflation target is strengthened after the fiscal policy is set, the leadership gains to the government will be less (Hughes Hallett and Viegi, 2002). This is an important point because it implies that granting leadership to a central bank whose inflation aversion is already greater than the government's, or whose commitment is greater, will produce no additional gains. For this reason, granting leadership to a central bank that cannot influence debt or deficit directly, but puts a higher priority on inflation control will bring smaller gains from a society's point of view (whatever it may do for the central bank). I demonstrate these points formally in Hughes Hallett (2005).

---

[23] $m_t \approx 0$; see Appendix B. This result still holds good when $\beta > \gamma$, even if monetary financing is allowed. It also holds good if the central bank is conservative, or the government is liberal, when $\beta < \gamma$.

Finally, given the structure of the Stackelberg game, there is no incentive to re-optimize for either party — so long as the government remains committed to long-term leadership rather than short-term demand management — unless both parties agree to reduce their independence through discretionary co-ordination.[24] This is a general result: it holds good for any model where both inflation and output depend on *both* monetary and fiscal policy, and where inflation is targeted to some degree by both players (Demertzis *et al.*, 2004; Hughes Hallett and Viegi, 2002). Thus, although Pareto improvements, over no co-operation, do not emerge for both parties in all Stackelberg games, they do so in problems of the kind examined here. Our results are robust.

6.1. *Empirical Evidence*

Whether or not these results are of practical importance is another matter. In Table 3, I have computed the expected losses under the simultaneous move and fiscal leadership regimes for 10 countries when both fiscal policy and the central bank are constructed optimally. The data I have used is from 1998, the year in which the Eurozone was created. The data itself, and its sources, are summarized in Appendix A.

The countries that are selected fall into three groups:

(a) Eurozone countries, with independently set fiscal and monetary policies and target independence: France, Germany, Italy, and the Netherlands;
(b) Explicit inflation targeters with fiscal leadership: Sweden, New Zealand, and the United Kingdom and
(c) Federalists, with joint inflation and stabilization objectives: United States, Canada, and Switzerland.

In the first group, monetary policy is conducted at the European level and fiscal policy at the national level. Policy interactions, in this group, can be characterized in terms of a simultaneous move game, with target as well as with instrument independence for both sets of policy authorities.

[24] This supplies the political economy of our results. We need no pre-commitment beyond leadership (the Stackelberg game supplies sub-game perfection, see Note 4) and no punishment beyond electoral results. The former will hold so long as governments have a natural commitment to maintaining sustainable public finances, public services (health, education, and defence), or social equity, all long-run "contractual" issues. The latter will then arise from the political competition inherent in any democracy.

Table 3. Expected losses under a deficit rule, leadership vs. other strategies.

	Full dependence $\delta = 1$ $\lambda^{cb} = \lambda_1^g$	Fiscal leadership $\delta = 0; \lambda_1^g = 1$ $\lambda^{cb} = 0$	Simultaneous moves $\delta = 0; \lambda_1^g = 1$ $\lambda^{cb} = \lambda^{cb*}$	Losses under simultaneous moves: Growth-rate equivalents
France	5.78	0.00	0.134	13.4
Germany	16.14	0.00	0.053	5.30
Italy	1.28	0.00	0.207	20.7
Netherlands	1.28	0.00	0.165	16.5
Sweden	4.51	0.00	0.034	3.40
New Zealand	8.40	0.00	0.071	7.10
United Kingdom	3.37	0.00	0.057	5.70
United States of America	6.46	0.00	0.248	24.8
Canada	12.50	0.00	0.118	11.8
Switzerland	4.79	0.00	0.066	6.60

It is, perhaps, arguable that the Eurozone has adopted a monetary leadership regime; but the empirical evidence for this was weak, and the fiscal constraints that could have sustained such a leadership were widely ignored.

The second group of countries has adopted explicit, and mostly publicly announced, inflation targets. Central banks in these countries have been granted instrument independence, but not target independence. The government either sets, or helps to set, the inflation target. In each case, the government has adopted long-term (supply side) fiscal policies, leaving an active demand management to monetary policy. These are clear cases in which there is both fiscal leadership and instrument independence for the central bank.

The third group represents a set of more flexible economies, with *implicit* inflation targeting and a statutory concern for stabilization; also federalism in fiscal policy and independence at the central bank, and therefore no declared leadership in either fiscal or monetary policies. This provides a useful benchmark for the other two groups.

The performance under the different fiscal regimes when the fiscal constraint is a relatively soft deficit rule is reported in Table 3. Column 1 shows the losses under a dependent central bank in welfare units — leadership or not. Column 2 reflects the losses that would be incurred under the government leadership with an independent central bank that directs monetary policy exclusively towards the achievement of the inflation target (i.e., $\delta = \lambda^{cb} = 0$). The third column gives the minimum loss associated with a simultaneous decision-making version of the same game. In each case, the deficit rule is soft for the Eurozone countries ($\lambda_2^g = 0.25$); medium strength for the United States, Canada, and Switzerland ($\lambda_2^g = 0.5$); and harder for Sweden, the United Kingdom and New Zealand ($\lambda_2^g = 1$). The deficit ratio targets are 0% in all countries (a balanced budget in the medium term), but the propensity to spend, s, out of any remaining deficit or any endogenous budget improvements is one.

Evidently, complete dependence in monetary policy is extremely unfavorable, although the magnitude of the loss varies considerably from country to country. The losses in Column 3 (Table 3) appear to be relatively small in comparison. However, when these figures are converted into "growth-rate equivalents", I find these losses to be significant. A growth-rate equivalent is the loss in output growth that would produce the same welfare loss, if all other variables remain fixed at their optimized values.[25]

The figures in Column 4 are more interesting. They show that the losses associated with simultaneous decision-making are equivalent to having permanent reductions of up to 20% in the level of national income. That is, France, Italy, and the Netherlands could have expected to have grown about 15%–20% more, had they adopted fiscal leadership with deficit targets; and Sweden, the United Kingdom, and New Zealand around 3%–5% less, had they not done so. Similarly, fiscal leadership with debt targeting would be worth 10%–20% extra GDP in the United States, and Canada. These are significant gains and significantly larger than could be expected from international policy co-ordination itself (Currie *et al.*, 1989).

Thus, the immediate effect of introducing a stability pact deficit rule has been to increase the costs of an unco-ordinated (simultaneous moves) regime dramatically. This reveals the weakness of the deficit rule as a

---

[25]Currie *et al.* (1989). To obtain these figures, I compute the marginal rates of transformation around each government's indifference curve to find the change in output growth, dy_t, that would yield the welfare losses in Column 3. Formally,

$$dy_t = \frac{dEL_t^g}{[\lambda_2^g\{(b-\theta)y_t - \tau_t\}(b-\theta) - \lambda_1^g]}$$

using Eq. (9). The minimum value of dy_t is, therefore, attained when tax revenues τ_t grow at the same rate as the re-payment target $(b-\theta)y_t$. These are the losses reported in Column 4.

restraint mechanism. Because it is a flow and not a stock, it has no inherent persistence. Moreover, a larger part (the *current* automatic stabilizers, social expenditures, and tax revenues) will be endogenous, varying with the level of economic activity and with the pressures of the electoral cycle. It is, therefore, much harder to use as a commitment device with any credibility in the long term. If we *assume* that it can be pre-committed, as in the leadership scenario, then we get good results of course. But, if this assumption is likely to be violated (or challenged), then the fact that the results are so much worse than in the simultaneous moves case shows how difficult it is to pre-commit deficits in advance. Being only a small component in a moving total, it does not have the persistence of a cumulated debt criterion because violations are not carried forward. So, when policy conflicts arise, the deficit either fails its target or it has to be restrained to such a degree that it starts to do serious damage to overall economic performance.

7. Conclusions

I conclude this chapter with the following observations:

(a) Fiscal leadership leads to improved outcomes because it implies a degree of co-ordination and reduced conflicts between institutions, without the central bank having to lose its ability to act independently. This places the outcomes somewhere between the superior (but discretionary) policies of complete co-operation; and the non-co-operative (or rule-based) policies of complete independence.
(b) The co-ordination gains come from the self-limiting action of fiscal policy when fiscal policy is given a long-run focus in the form of a (soft) debt or deficit rule. Leadership is the crucial element here; these gains do not appear in a straight-forward competitive regime with the same debt or deficit rule.
(c) The leadership model predicts improvements in inflation and the fiscal targets, without any loss in growth or output volatility. In addition, leadership requires less precision in the setting of the strategic or institutional parameters. It is easier to implement.
(d) Debt ratios will increase in a policy game without co-ordination, but do not do so under fiscal leadership. Likewise, deficit rules without co-ordination to restrain fiscal policy imply larger deficits (and more expansionary policies) than do debt rules of a similar specification because deficits (being a flow not a stock, and with less natural persistence) are less easy to pre-commit credibly. These two results explain

why the ECB prefers hard restraints (and why the fiscal policy makers do not), and suggests that the ECB may actually prefer fiscal leadership because of the greater degree of pre-commitment implied.

References

Barro, RJ (1981). Output effects of government purchases. *Journal of Political Economy* **89**, 1086–1121.

Barro, RJ and DB Gordon (1983). Rules, discretion, and reputation in a model of monetary policy. *Journal of Monetary Economics* **12**, 101–121.

Basar, T (1989). Time consistency and robustness of equilibria in non-cooperative dynamic games. In *Dynamic Policy Games in Economics*, F van der Ploeg and A de Zeeuw (eds.), pp. 9–54. Amsterdam: North Holland.

Basar, T and G Olsder (1989). Dynamic noncooperative game theory. In *Classics in Applied Mathematics*, SIAM series, Philadelphia: SIAM.

Brandsma, A and A Hughes Hallett (1984). Economic conflict and the solution of dynamic games. *European Economic Review* **26**, 13–32.

Buti, M, S Eijffinger and D Franco (2003). Revisiting the stability and growth pact: grand design or internal adjustment? *Discussion Paper 3692*. London: Centre for Economic Policy Research.

Canzoneri, M (2004). A New View on The Transatlantic Transmission of Fiscal Policy and Macroeconomic Policy Coordination. In *The Transatlantic Transmission of Fiscal Policy and Macroeconomic Policy Coordination*, M Buti (ed.), Cambridge: Cambridge University Press.

Currie, DA, G Holtham and A Hughes Hallett (1989). The theory and practice of international economic policy coordination: does coordination pay? In *Macroeconomic Policies in an Interdependent World*, R Bryant, D Currie, J Frenkel, P Masson and R Portes (eds.), Washington, DC: International Monetary Fund, 125–148.

Demertzis, M, A Hughes Hallett and N Yiegi (2004). An Independent Central Bank Faced with Elected Governments: A Political Economy Conflict. In *European Journal of Political Economy* **20**(4), 907–922.

Dieppe, A, K Kuster and P McAdam (2004). Optimal monetary policy rules for the Euro Area: an analysis using the area wide model. *Working Paper 360*. Frankfurt: European Central Bank.

Dixit, A (2001). Games of monetary and fiscal interactions in the EMU. *European Economic Review* **45**, 589–613.

Dixit, AK and L Lambertini (2003). Interactions of commitment and discretion in monetary and fiscal issues. *American Economic Review* **93**, 1522–1542.

Dow, JCR (1964). *The Management of the British Economy 1945–1960*. Cambridge: Cambridge University Press.

European Commission (2002). Public finances in EMU:2002. *European Economy: Studies and Reports* **4**, Brussels: European Commission.

Fatas, A, J von Hagen, A Hughes Hallett, R Strauch and A Sibert (2003). Stability and Growth in Europe. London: Centre for Economic Policy Research (MEI-13).

Gali, J and R Perotti (2003). Fiscal policy and monetary integration in Europe. *Economic Policy* **37**, 533–571.

HM Treasury (HMT) (2003). Fiscal Stabilization and EMU. HM Stationary Office, Cmnd 799373. Norwich: also available from www.hm-treasury.gov.uk.

Hughes Hallett, A (1984). Noncooperative strategies for dynamic policy games and the problem of time inconsistency. *Oxford Economic Papers* **36**, 381–399.

Hughes Hallett, A (2005). In praise of fiscal restraint and debt rules: what the Euro zone might do now. *Discussion Paper 5043*. London: Centre for Economic Research, 251–279.

Hughes Hallett, A and D Weymark (2002). Government leadership and central bank design. *Discussion Paper 3395*, London: Centre for Economic Research.

Hughes Hallett, A and N Viegi (2002). Inflation targeting as a coordination device. *Open Economies Review* **13**, 341–362.

Hughes Hallett, A and D Weymark (2004a). Policy games and the optimal design of central banks. In *Money Matters*, P Minford (ed.). London: Edward Elgar.

Hughes Hallett, A and D Weymark (2004b). Independent monetary policies and social equity. *Economics Letters* **85**, 103–110.

Hughes Hallett, A and D Weymark (2005). Independence before conservatism: transparency, politics and central bank design. *German Economic Review* **6**, 1–25.

Lucas, RE (1972). Expectations and the neutrality of money. *Journal of Economic Theory* **4**, 103–124.

Lucas, RE (1973). Some international evidence on output-inflation trade-offs. *American Economic Review* **63**, 326–334.

McCallum, BT (1989). *Monetary Economics: Theory and Policy*. New York: Macmillan.

Svensson, L (2003). What is wrong with Taylor rules? Using judgement in monetary rules through targeting rules. *Journal of Economic Literature* **41**, 426–477.

Taylor, JB (1993a). *Macroeconomic Policy in a World Economy: From Econometric Design to Practical Operation*. New York: W.W. Norton and Company.

Taylor, JB (1993b). Discretion vs. policy rules in practice. *Carnegie-Rochester Conference Series on Public Policy* **39**, 195–214.

Taylor, JB (2000). Discretionary fiscal policies. *Journal of Economic Perspectives* **14**, 1–23.

Turrini, A and J in 't Veld (2004). The Impact of the EU Fiscal Framework on Economic Activity. DGII, Brussels: European Commission.

Appendix A: Data Sources for Table 3

The data used in Table 3 are set out in the table below. They come from different sources and represent "stylized facts" for the corresponding parameters. The Phillips curve parameter, α, is the inverse of the annualized sacrifice ratios on quarterly data from 1971–1998. The values for β and γ, the impact multipliers for fiscal and monetary policy, respectively, are the one-year simulated multipliers for these policies in Taylor's multicountry model (Taylor, 1993a). I have calibrated the remaining parameters using stylized facts from 1998: the year that EMU started, and the year that new fiscal and monetary regimes took effect in the United Kingdom and Sweden. Thus, s is the automatic stablizer effect on output according to the European Commission (2002) and HMT (2003) estimates, allowing for larger social

Table 4. Data for the deficit rule calculations.

	α	β	γ	s	θ	ϕ	λ_2^g
France	0.294	0.500	0.570	1	0	1.147	0.25
Germany	0.176	0.533	0.430	1	0	1.094	0.25
Italy	0.625	0.433	0.600	1	0	1.271	0.25
Netherlands	0.625	0.489	0.533	1	0	1.306	0.25
Sweden	0.333	0.489	0.533	1	0	1.163	1
New Zealand	0.244	0.400	0.850	1	0	1.098	1
UK	0.385	0.133	0.580	1	0	1.038	1
United States	0.278	0.467	1.150	1	0	1.130	0.5
Canada	0.200	0.400	0.850	1	0	1.080	0.5
Switzerland	0.323	0.489	0.533	1	0	1.158	0.5

security sectors in the Netherlands and Sweden. Similarly, θ is set to give a debt target somewhat below each country's 1998 debt ratio and the SGP's 60% limit. Likewise, λ_2^g was set to imply a low, high, or medium commitment to debt/deficit limits, as reflected in actual behavior in 2004. Finally, I have set $\lambda_1^g = 1$ in each country to suggest that governments, if left to themselves, are equally concerned about inflation and output stabilization.

For Table 4, I suppose that the deficit that remains after growth effects and new discretionary taxes are accounted for will be spent ($s = 1$), and that the medium-term deficit target is zero ($\theta = 0$).

Appendix B: Derivation of the Change in Fiscal Stance Between Regimes

The expected value of $s(by_t - \tau_t)$ under fiscal leadership and simultaneous moves can be obtained by inserting values first from Eq. (29), and then from Eqs. (31) and (32), respectively. Taking the expected values, this yields zero and $\phi\lambda^{cb*}/(\alpha^2\gamma) > 0$ in each case. Applying this result to Eq. (4), it shows that the *expected* change in output between regimes, which we know to be zero, is $(\beta - \gamma)\Delta m - \beta\lambda^{cb*}/(\alpha\lambda_1^g) + \gamma\Delta[s(by - \tau)] = 0$. Hence,

$$\Delta m = \{\gamma\Delta[s(by - \tau)] + \beta\lambda^{cb^*}/(\alpha\lambda_1^g)\}/(\beta - \gamma)$$

which, given the change in $s(by_t - \tau_t)$, is positive, if $\beta > \gamma$. In this case, the simultaneous move regime is unambiguously more expansionary, and will run larger deficits, than a regime with fiscal leadership. But, since

$\gamma/(\beta - \gamma) > 1$, this result may not follow if $\beta < \gamma$. In the latter case, the expansion of the budget deficit will be:

$$\begin{aligned}\Delta Deficit &= \Delta m + \Delta[s(by - \tau)] - \Delta\tau \\ &= \left[\frac{\beta}{\alpha(\beta - \gamma)}\left(\frac{\phi}{\alpha\gamma} + \frac{1}{\lambda_1^g}\right) - \frac{\gamma s}{\phi\lambda_2^g}\right]\lambda^{cb*} + \frac{\gamma s}{\varphi\lambda_2^g}\lambda_1^g\end{aligned}$$

which is still clearly positive if *either* λ^{cb^*} is small, the central bank is conservative, *or* if λ_1^g is somewhat larger (the government is relatively liberal).

Topic 2

Cheap-Talk Multiple Equilibria and Pareto — Improvement in an Environmental Taxation Games

Chirstophe Deissenberg

Université de la Méditerranée and GREQAM, France

Pavel Ševčík

Universilé de Montréal, Canada

1. Introduction

The use of taxes as regulatory instrument in environmental economics is a classic topic. In a nutshell, the need for regulation usually arises because producing causes detrimental emissions. Due to the lack of a proper market, the firms do not internalize the impact of these emissions on the utility of other agents. Thus, they take their decisions on the basis of prices that do not reflect the true social costs of their production. Taxes can be used to modify the prices, confronting the firms so that the socially desirable decisions are taken. The problem has been exhaustively investigated in static settings, where there is no room for strategic interaction between the regulator and the firms.

Consider, however, the following situations:

(1) The emission taxes have a dual effect, they incite the firms to reduce production and to undertake investments in abatement technology. This is typically the case, when the emissions are increasing in the output and decreasing in the abatement technology;
(2) Emission reduction is socially desirable. The reduction of production is not and
(3) The investments are irreversible. In this case, the regulator must find an optimal compromise between implementing high taxes to motivate high investments, and keeping the taxes low to encourage production.

The fact that the investments are irreversible introduces a strategic element in the problem. If the firms are naïve and believe his announcements, the regulatory can insure high production and important investments by first declaring high taxes and reducing them, once the corresponding investments have been realized. More sophisticated firms, however, recognize that the initially high taxes will not be implemented, and are reluctant to invest in the first place. In other words, one is confronted with a typical time inconsistency problem, which has been extensively treated in the monetary policy literature following Kydland and Prescott (1977) and Barro and Gordon (1983). In environmental economics, the time inconsistency problem has received yet only limited attention, although it frequently occurs. See among others Gersbach and Glazer (1999) for a number of examples and for an interesting model, see Abrego and Perroni (1999); Batabyal (1996a;b); Dijkstra (2002); Marsiliani and Renstrom (2000) and Petrakis and Xepapadeas (2003).

The time inconsistency is directly related to the fact that the situation described above, defines a Stackelberg game between the regulator (the leader) and the firms (the followers). As noted in the seminal work of Simaan and Cruz (1973a;b), inconsistency arises because the Stackelberg equilibrium is not defined by mutual best responses. It implies that the follower uses a best response in reaction to the leader's action, but not that the leader's actions is itself a best response to the follower's. This opens the door to a re-optimization by the leader once the follower has played. Thus, a regulator, who announces that it will implement the Stackelberg solution, is not credible. An usual conclusion is that, in the absence of additional mechanisms, the economy is doomed to converge towards the less desirable Nash solution.

A number of options to insure credible solutions have been considered in the literature — credible binding commitments by the Stackelberg leader, reputation building, use of trigger strategies by the followers, etc. (see McCallum (1997), for a review in a monetary policy context). Schematically, these solutions aim at assuring the time consistency of Stackelberg solution with either the regulator or the firms as a leader. Usually, these solutions are not efficient and can be Pareto-improved.

In Dawid *et al.* (2005), we demonstrated the usefulness of a new solution to the time inconsistency problem in environmental policy. We showed that tax announcements, that are not respected, can increase the payoff not only of the regulator, but also of all firms, if these include any number of naïve believers who take the announcements at face value. Moreover, if firms tend to adopt the behavior of the most successful ones, a unique stable equilibrium may exist where a positive fraction of firms are believers. This equilibrium Pareto-dominates the one, where all firms anticipate perfectly the Regulator's action. To attain the superior equilibrium, the regulator builds reputation and leadership by making announcements and implementing taxes in a way that generates good results for the believers, rather than by pre-committing to his announcements.

The potential usefulness of employing misleading announcements to Pareto-improve, upon standard game-theoretic equilibrium solutions was suggested for the case of general linear-quadratic dynamic games in Vallee *et al.* (1999) and developed by the same authors in subsequent papers. An early application to environmental economies is Vallee (1998). The believers/non-believers dichotomy was introduced by Deissenberg and Gonzalez (2002), who study the credibility problem in monetary economics in a discrete-time framework with reinforcement learning. A similar monetary policy problem has been investigated by Dawid and Deissenberg (2005) in a continuous-time setting akin to the used in the present work.

In this chapter, we re-visit the model proposed by Dawid *et al.* (2005) and show that a minor (and plausible) modification of the regulator's objective function, opens the door to the existence of multiple stable equilibria with distinct basins of attraction — with important interpretations and potential consequences for the practical conduct of environmental policy.

This chapter is organized as follows. In Section 2, we present the model of environmental taxation and introduce the imitation-type dynamics that determinate the evolution of the number to believers in the economy. In Section 3, we derive and discuss the solution of the sequential Nash game, one obtains by assuming a constant proportion of believers. In Section 4, we formulate the dynamic problem and investigate its solution numerically. Finally, in Section 6, we summarize the mechanisms at work in the model and the main insights.

2. The Model

We consider an economy, consisting of a Regulator R and a continuum of atomistic, profit-maximizing firms i with an identical production technology. Time τ is continuous. To keep the notation simple, we do not index the variables with either i or r unless it is useful for a better understanding.

In a nutshell, the situation of interest is the following: The regulator can tax the firms to incite them to reduce their emission, and to generate tax income, taxes, however, have a negative impact on the employment (and thus, on the labor income) and the profits. Thus, R has to choose the tax level in order to achieve an optimal compromise among tax revenue, private sector income, emission reduction, and employment levels. The following sequence of event (S) occurs in every τ:

- R makes a non-binding announcement $t^a \geq 0$ about the tax level $t \geq 0$ it will choose.
- Given t^a, the firms form expectations t^e about the true level of the environmental tax. As will be described in more detail later, there are two different ways for an individual firm to form its expectations.
- Each firm decides about its level of investment v based on its expectation t^e.
- R choose the actual level of tax $t \geq 0$. The tax level t is the amount each firm has to pay per unit of its emissions.
- Each firm produces a quantity x.
- Each firm may revise the way it forms its expectations depending on its relative profits.

All firms are identical, except for the way they form their expectations t^e. We assume full information: the regulator and the firms perfectly know both the sequence, we just presented and the economic structure will be (production and prediction technology, objective functions ...) described in the remainder of this section.

2.1. *The Firms, B*s *and NB*s

Each firm produces the same homogenous goods using a linear production technology: the production of x units of output, requires x units of labor and generates x units of environmentally damaging emissions. The production costs are given by:

$$c(x) = wx + c_x x^2, \tag{1}$$

where x is the output, $w > 0$ the fixed wage rate, and $c_x > 0$ a parameter for simplicity sake, the demand is assumed infinitely elastic at the given price $\tilde{p} > w$. Let $p := \tilde{p} - w > 0$. At each point of time, each firm can spend an additional amount of money γ in order to reduce its current emissions x. The investment

$$\gamma(v) = \frac{1}{2}v^2 \tag{2}$$

is needed, in order to reduce the firm's emissions from x to $\max\,[x - v, 0]$. The investment in one period has no impact on the emissions in future period. Rather than expenditures in emission-reducing capital, γ can therefore be interpreted as the additional costs resulting of a temporary switch to a cleaner resource — for e.g., of a switch from coal to natural gas.

The firms attempts to maximize their profit. As we shall see at a later place, one can assume without loss of generality that they disregard the future and maximize in every τ, their current *instantaneous* profit. However, the actual tax rate t enters their instantaneous profit function and is unknown at the time when they choose v. Thus, they need to base their optimal choice of v on an expectation (or, prediction) t^e of t. In reality, a firm can invest larger or smaller amounts of money in order to make or less accurate predictions of the tax, the government will actually implement. In this model, we assume for simplicity that each firm has two (extreme) options for predicting the implemented tax. It can either suppose that R's announcement is truthful, that is, that t will be equal to t^a, or it can do costly research and try to better predict the t that will actually be implemented. Depending upon the expectation-formation mechanism the firm chooses, we speak of believer B or of a non-believer NB:

A believer considers the regulator's announcement to be truthful and sets

$$t^e = t^a \tag{3}$$

This does not cost anything.

A non-believer makes a "rational" prediction of t, considering the current number of Bs and NBs as constant. Making the prediction in any given period τ costs $\delta > 0$.

Details on the derivation of the NBs' prediction will be given later. Note that a firm can change its choice of expectation-formation mechanism at any moment in time. That is, a B can become a NB, and *vice versa*. We denote with $\pi = \pi(t) \in [0, 1]$ the current fraction of Bs in the population. Thus, $1 - \pi$ is the current fraction of NBs.

2.2. *The Regulator, R*

The regulator's goal is to maximize with respect to $t^a \geq 0$ and $t \geq 0$, the inter-temporal social welfare function.

$$\Phi(t^a, t) = \int_0^\infty e^{-\rho\tau} \varphi(t^a, t) \mathrm{d}\tau$$

with

$$\varphi(t^a, t) = (\pi x^B + (1-\pi)x^{NB}) + \pi g^B + (1-\pi)g^{NB}$$
$$- (\pi(x^B - v^B) + (1-\pi)(x^{NB} - v^{NB}))^2$$
$$+ t(\pi(x^B - v^B) + (1-\pi)(x^{NB} - v^{NB})), \tag{4}$$

where x^B, x^{NB}, v^B, v^{NB} denote the production respectively, investment chosen by the believers B and the non-believers NB, and where g^B (.) and g^{NB} (.) are their instantaneous profits,

$$g^B = px^B - c_x(x^B)^2 - t(x^B - v^B) - \frac{1}{2}(v^B)^2$$
$$g^{NB} = px^{NB} - c_x(x^{NB})^2 - t(x^{NB} - v^{NB}) - \frac{1}{2}(v^{NB})^2 - \delta. \tag{5}$$

The strictly positive parameter ρ is a social discount factor.

Thus, the instantaneous welfare ϕ is the sum of: (1) the economy's output, that is also, the level of employment (remember that output and employment are one-to-one in this economy); (2) the total profits of *B*s and *NB*s; (3) the squared volume of emissions and (4) the tax revenue. This function can be recognized as a reasonable approximation of the economy's social surplus. The difference between the results of Dawid *et al.* (2005) and those of the present article stem directly from the different specification of the regulator's objective function. In the former article, this function does not include the firm's profit and is linear in the emissions.

2.3. *Dynamics*

The firms switch between the two possible expectation-formation mechanisms (*B* or *NB*), according to a imitation-type dynamics, see Dawid (1999); Hofbauer and Sigmund (1998). Specifically, at each τ the firms meet randomly two-by-two, each pairing being equiprobable. At every encounter, the firm with the lower current profit adopts the belief of the other firm with a probability proportional to the current difference between the individual profits. Thus, on the average, the firms tend to adopt the type of prediction, *N* or *NB*, which may currently leads to the highest profits. Above mechanism gives rise to the dynamics.

$$\frac{d\pi}{dt} = \dot{\pi} = \beta\pi(1-\pi)g^B - g^{NB}. \tag{6}$$

Notice that $\dot{\pi}$ reaches its maximum for $\pi = \frac{1}{2}$ (the value or π for which the probability, for encounter between firms with different profits are maximized), and tends towards 0, for $\pi \to 0$ and $\pi \to 1$ (for extreme values of π, all but very few firms have the same profits). The parameter $\beta \geq 0$, that captures the probability with which a firm changes its strategy during one of these meetings, calibrates the speed of the information flow between *B*s and *NB*s. It can be interpreted as a measure of the firms' willingness to change strategies, that is, of the flexibility of the firms.

Equation (6) implies that by choosing the value of (t^a, t) at time τ the regulator not only influences the instantaneous social welfare, but also the future proportion of *B*s in the economy. This, in turn, has an impact on the future social welfare. Hence, there are no explicit dynamics for the economy. This again has an impact on the future social welfare. Hence, although there are no explicit dynamics for the economic variables v and x, R faces a non-trivial inter-temporal optimization problem.

On the other hand, since the firms are atomistic, each single producer is too small to influence the dynamics Eq. (6) of π, the only source of dynamics in the model. Thus, the single firm does not take into account any inter-temporal effect and, independently of its true planing horizon, maximizes its current profit in every τ. This naturally leads us to examine the (Nash) solution of the game between the regulator and the non-believers that follows from the sequence (S) when, for an arbitrary, fixed value of π, (a) R maximizes the integrand ϕ in Eq. (4) with respect to t^a and t and (b) the *NB*s maximize their profits with respect to v^{NB} and x^{NB}. As previously mentioned, we assume full information. In particular, the R and the *NB*s are fully aware of the (non-strategic) behavior of the *NB*s.

The analysis of the static case is also necessary since, as previously mentioned, it provides the prediction t^e of the *NB*s.

3. The Sequential Nash Game

The game is solved by backwards induction, taking into account the fact that the firms, being atomistic, rightly assume that their individual actions do not affect the aggregate values, that is consider these values as given. We consider only symmetric solutions where all *B*s respectively, *NB*s choose the same values for v and x.

3.1. *Derivation of the Solution*

The last decision in the sequence (S) is the choice of the production level x by the firms. This choice is made after v and t are known. The firms being takers, the optimal symmetric production decision is:

$$x^B = x^{NB} = x = \frac{p - t}{2c_x}. \tag{7}$$

Note that the optimal production is the same for Bs and NBs, as it exclusively depends upon the *realized* taxes t.

In the previous stage, R chooses t given v^B. Maximizing v^{NB} with respect to t, with x is given by (7) gives the optimal reaction function:

$$t(v^B, v^{NB}) = \frac{p - c_x[2(\pi v^B + (1-\pi)v^{NB}) + 1]}{1 + c_x}. \tag{8}$$

When the firms determine their investment v, Eqs. (7) and (8), and t^a are known, but the actual tax rate t is not. Using Eq. (7) in Eq. (5), one recognizes that the firms choose v in order to maximize

$$\frac{(p - t^e)^2}{4c_x} + t^e v - \frac{1}{2}v^2. \tag{9}$$

That is, they set:

$$v = t^e \tag{10}$$

The Bs predict that t will be equal to its announced value, $t^e = t^a$. Thus,

$$v^B = t^a. \tag{11}$$

The NBs know that the regulator will act according to Eq. (8). Thus, they choose:

$$v^{NB} = \frac{p - c_x[2(\pi v^B + (1-\pi)v^{NB}) + 1]}{1 + c_x}. \tag{12}$$

Solving this last equation for v^{NB} gives for the equilibrium value:

$$v^{NB} = v^{NB}(t^a)\frac{p - c_x(1 + 2\pi t^a)}{1 + c_x(3 - 2\pi)}. \tag{13}$$

The first decision to be taken is the choice of t^a by R. The regulator's instantaneous objective function ϕ, using the results from the previous stages, i.e., Eq. (7) for x, Eq. (11) for v^B and Eq. (13) for v^{NB}, is concave with respect to t^a for all $\pi \in [0, 1]$, if,

$$c_x > 1/\sqrt{3}. \tag{14}$$

We shall assume in the following that this inequality holds. Then, the regulator will choose

$$t^{a\$} = \frac{1+p}{1+3c_x}. \tag{15}$$

Using this last equation in Eq. (8) with v^{NB} given by Eq. (13), one obtains for the equilibrium value of t

$$t^{\$} = \frac{1+p}{1+3c_x} - \frac{1+c_x}{c_x(3-2\pi)+1}. \tag{16}$$

Both $t^{a\$}$ and $t^{\$}$ will be non-negative if

$$p \geq \frac{3}{3c_x - 2}, \tag{17}$$

which we assume from now on.

3.2. *Some Properties of the Solution*

For all $c_x > 0$ and $\pi \in [0, 1]$, the optimal announcement $t^{a\$}$ is a constant, and always less than $t^{a\$}$. It is optimal to announce a high tax in order to elicit a high investment from the believers, and to implement a low tax to motivate a high production level. Moreover, $t^{\$}$ decreases in π: When the proportion of *B*s augments, the announcement t^a becomes a more powerful instrument making a recourse to t less necessary.

The difference between the statically optimal (i.e., evaluated at $t^{\$}$) profits of *NB*s and *B*s

$$g^{NB\$} - g^{B\$} = \frac{(1+c_x)^2}{2c_x(2\pi-3)^2} - \delta, \tag{18}$$

is likewise increasing in π. Modulo δ, the profit of the *NB*s is always higher than the profit of the *B*s whenever $\pi > 0$, reflecting the fact that the latter make a systematic error about the true value of t. The profit of the *B*s, however, can exceed the one of the *NB*s if the prediction costs δ are high enough.

Since $t^{a\$}$ is constant and $t^{\$}$ decreasing in π and since $t^{\$} \leq t^{a\$}$, the difference $t^{a\$} - t^{\$}$, increases with π. Therefore, the difference Eq. (18) between the profits of the *NB*s and *B*s is also increasing in the difference between the announced and the implemented tax, $t^{a\$} - t^{\$}$,

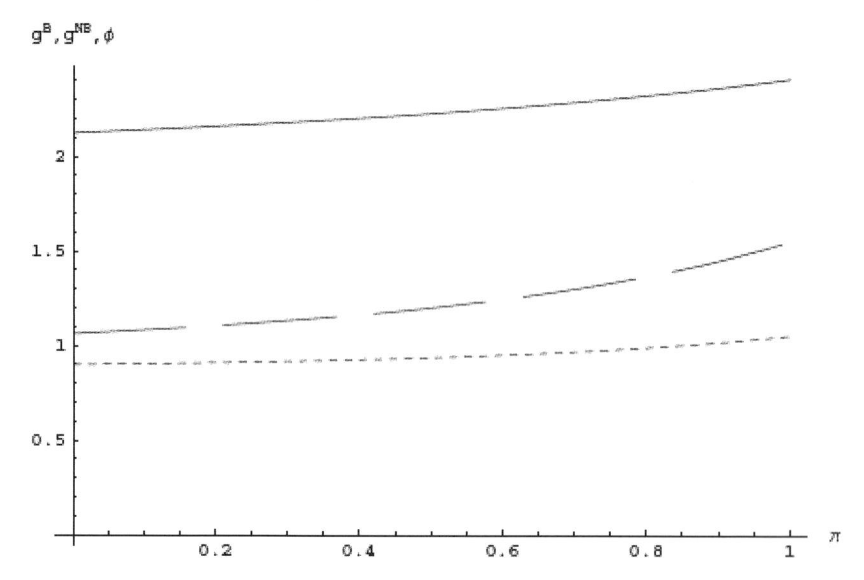

Figure 1. The profits of the *B*s (lower curve), of the *NB*s (middle curve), and the regulator's utility (upper curve) as a function of π for $\delta = 0.00025, c_x = 0.6, p = 2$.

As one would intuitively expect, the regulator's instantaneous utility ϕ too, increases with π. Most remarkable, however, is the fact that the profits of both the *B*s and *NB*s are also increasing in π (see Fig. 1 for an illustration). An intuitive explanation is the following: For $\pi = 0$ the outcome is the Nash solution of a game between the *NB*s and the regulator. This outcome is not efficient leaving room for Pareto-improvement. As π increases, the problem tends towards the solution a standard *optimization* situation with the regulator as the sole decision-maker. This solution is *efficient* in the sense that it does maximize the regulator's objective function under the given constraints. Since the objectives of the regulator are not extremely different of the objectives of the firm, moving from the inefficient game-theoretic solution to the efficient control-theoretic solution one improves the position of all actors. Specifically, as π increases, the *B*s enjoy the benefits of a lower t, while the investment remains fixed at $t^{a\$}$. Likewise, the *NB*s benefit from the decrease in t. And R also benefits, because a raising number of *B*s are led by R's announcement to invest more than they would, otherwise and, subsequently, to produce more, to make higher profits, and to pollute subsequently, to produce more, to make higher profits, and to pollute less.

In dynamic settings, however, the impact of the tax announcements on the π dynamics may incite the regulator to reduce the spread between t and t^a, and in particular to choose $t^a < t^{a\$}$. *Ceteris paribus*, the R prefers a high proportion of *B*s to a low one, since it has one more instrument (namely, t^a) to influence the *B*s than to regulate the *NB*s. A high spread $t^a - t$, however, implies that the profits of the *B*s will be low compared to those of the *NB*s. This, by Eq. (6), reduces the value of $\dot{\pi}$ and leads over time to a lower proportion of *B*s, diminishing the instantaneous utility of *R*. Therefore, in the dynamic problem, *R* will have to find an optimal compromise between choosing a high value of t^a, which allows a low value of t and insures the regulator a high instantaneous utility, and choosing a lower one, leading to a more favorable proportion of *B*s in the future. We are going to explore these mechanisms in more details in the next section.

4. The Dynamic Optimization Problem

4.1. *Formulation*

Being atomistic, the firms do not recognize that their individual decisions to choose the *B* or *NB* prediction technology influence π over time. Assume that they consider π as constant. The regulator's optimization problem is then:

$$\begin{aligned}&\max \Phi(t^a, t) \quad 0 \leq t^a(\tau), t(\tau)\\&\text{such that Eqs. (6), (7), (11), (13),} \quad \text{and} \quad \pi_0 = \pi(0).\end{aligned} \tag{19}$$

Using

$$\begin{aligned}g &:= \frac{p - c_x(1 + 2\pi t^a)}{1 + c_x(3 - 2\pi)},\\ h &:= \frac{p - t}{2c_x}.\end{aligned} \tag{20}$$

The Hamiltonian of the above problem is given by

$$\begin{aligned}H(t^a, t, \pi, \lambda) = h &+ \pi\left[h - \frac{(p-t)^2}{4c_x} + t(h - t^a) - \frac{t^{a2}}{2}\right]\\&+ (1-\pi)\left[h - \frac{(p-t)^2}{4c_x} + t(h - g) - \frac{1}{2}g^2 - \delta\right]\\&+ t[\pi(h - t^a) + (1-\pi)(h - g)]\\&- [\pi(h - t^a) + (1-\pi)(h-g)]^2\\&+ \lambda\pi(1-\pi)\beta\left[tt^a - \frac{t^{a2}}{2} - th + \frac{1}{2}h^2 + \delta\right],\end{aligned}$$

where λ is the co-state.

The first order maximization condition with respect to t^a and t yields for the dynamically optimal values*

$$t^{a*} = \frac{(p - c_x) + [c_x(3 - 2\pi) + 1](1 + c_x)^2}{(1 + 3c_x)[\beta\lambda(1 - \pi) + K]}, \tag{21}$$

$$t^* = \frac{(p - c_x) + 2c_x(1 + c_x)\pi[c_x(\beta\lambda(1 + 3c_x)(1 - \pi) + 1)]}{(1 + 3c_x)[\beta\pi(1 - \pi) + K]}$$

with

$$K = 1 - 6c_x^3\beta^2\lambda^2(1 - \pi)^2\pi + 2c_x[2 + 2\beta\lambda(1 - \pi)\pi] \\ + c_x^2[3 - 2\pi + \beta\lambda(1 - \pi)(3 - 2\beta\lambda(1 - \pi)\pi)]. \tag{22}$$

According to Pontriagin's maximum principle, an optimal solution $(\pi(\tau), (\lambda))$ has to satisfy the state dynamics (6), evaluated at (t^{a*}, t^*) plus:

$$\dot{\lambda} = \rho\lambda - \frac{\partial H(t^{a*}(\pi, \lambda), t^*(\pi, \lambda), \pi, \lambda)}{\partial \pi} \tag{23}$$

$$0 = \lim_{\tau \to \infty} e^{-\rho\tau}\lambda(\tau). \tag{24}$$

Due to the highly non-linear structure of the optimization problem an explicit derivation of the optimal $(\pi(\tau), \lambda(\tau))$ seems infeasible. Hence, we used Mathematica® to carry out a numerical analysis of the canonical system Eqs. (6)–(23) in order to shed light on the optimal solution.

4.2. *Numerical Results*

Depending upon the parameter values, the model can have either two stable equilibria separated by a so-called Skiba point, or a unique stable equilibrium. We discuss first the more interesting case of multiple equilibria, before carrying a summary of sensitivity analysis and briefly addressing the question of the bifurcations, leading to a situation with a situation with unique equilibrium. Unless and otherwise specified, the results presented pertain to the reference parameter configuration $c_x = 0.6$, $p = 2$, $\delta = 0.00025$ that underlies Fig. 1, together with $\rho = 1, 0.0015$. These are robust, with respect to sufficiently small parameter variations: The same qualitative insights would be obtained for a compact set of alternative parameter configurations.

From the onset, let us stress the fundamental mechanism that underlies most of the results. *Ceteris paribus*, R would like to face as many Bs as possible, since $\mathrm{d}\phi/\mathrm{d}\pi > 0$. If the prediction costs δ are sufficiently high, the profit difference $g^{NB\$} - g^{B\$}$ see Eq. (18), will be negative at the current value of π. In this case, by Eq. (6), implementing the statically optimal actions $t^{a\$}$, $t^{\$}$ suffices (*albeit* not optimally so) to insure that π increases locally. In fact, for δ sufficiently large, the thus controlled system will globally converge to a situation where there are only believers, $\pi = 1$. However, if δ is small, the regulator cannot implement the actions $t^{a\$}$, $t^{\$}$ since in that case $g^{NB} < g^{B}$, which implies by Eq. (6) a decreasing number of believers, $\dot{\pi} < 0$. Instead, R will use actions t^{a*}, t^{*} that insure an optimal compromise between obtaining as high a value of π and the optimal value $\dot{\pi}^{*}$ of $\dot{\pi}$ may be negative. In this case, the economy will tend towards the already discussed equilibrium, $\$$ where nobody believes R, $\pi = 0$. At this equilibrium, everybody correctly anticipates the tax level $t^{\$}$. However, the outcome is inefficient.

Alternatively, $\dot{\pi}^{*}$ may be positive. Then, there will be a positive number of *B*s at the equilibrium. Since for any given y^{a}, y the speed of learning $\dot{\pi}$ is first increasing and then decreasing in π, see Eq. (6), it will not come as a surprise, if for certain parameter constellations there exists a value of π that defines a threshold between the two domains of convergence, one towards $\pi = 0$ and one towards some $\pi u > 0$.

4.3. *Multiple Equilibria*

4.3.1. *Phase Diagram*

The phase diagram for the reference parameter values, $c_{x} = 0.6$, $p = 2$, $\delta = 0.00025$, $\beta = 1$, and $\rho = 0.0015$, is shown in Fig. 2.

The $\dot{\pi} = 0$ isocline, shown in dotted lines, has three branches. Two of these correspond to the situation where there is a homogenous population of *NB*s ($\pi = 0$) or of *B*s ($\pi = 1$). Along the third branch, *B*s and *NB*s co-exist but make the same prediction errors and realize the same profits and decisions. The $\dot{\lambda} = 0$ isocline is shown as a continuous line. There are three equilibria, $E_{L} = (0, \lambda_{L})$, $E_{M} = (\pi_{M}, \lambda_{M})$, and $E_{U} = (\pi_{U}, \lambda_{U})$ with $0 < \pi_{M} < \pi_{U} < 1$. The lower and upper equilibrium, E_{L} and E_{U}, are locally stable. The middle equilibrium E_{M} is an unstable spiral.

Specifically, for the reference parameter values, $E_{M} = (0.24, 5.98)$ and $E_{U} = (0.93, 16.60)$. The characteristic roots of the Jacobian of the Canonical system Eqs. (6), (23), evaluated at E_{M}, E_{U}, respectively are 0.04 and −0.03 (indicating saddle-point stability), respectively $0.0075 \pm 0.0132i$ (characteristic for an unstable spiral). Most interestingly, E_{U} Pareto-dominates E_{L}, although at the latter equilibrium all firms make perfect predications of the taxes, the regulator is going to implement.

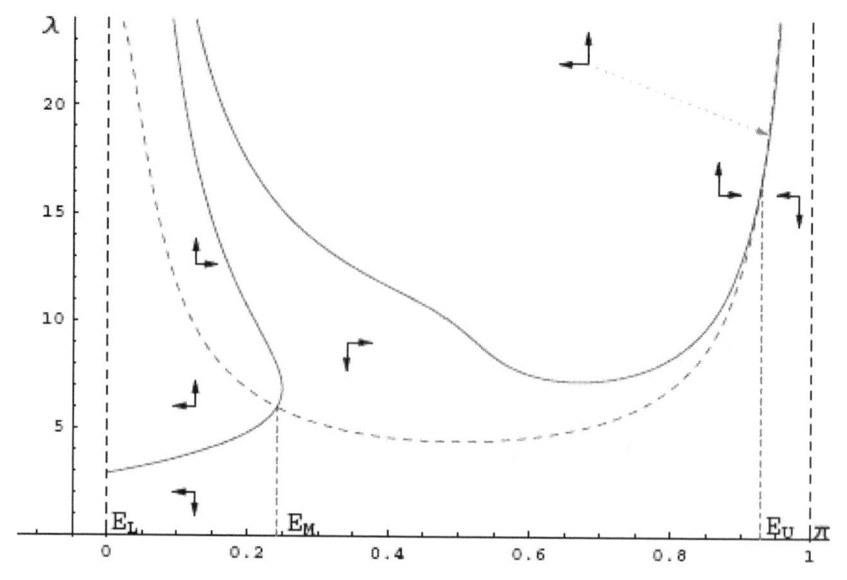

Figure 2. Phase diagram of the canonical system, $c_x = 0.6$, $p = 2$, $\delta = 0.00025$, $\beta = 1$, and $\rho = 0.0015$.

Table 1. Profits, welfare, and taxes at E_U and $E_L, c_x = 0.6, p = 2, \delta = 0.0025, \beta = 1, p = 0.0015$.

	g	ϕ	t^a	t
E_U	1	2.25	1.07	0.12
E_L	0.98	2.18	N.A.	0.62

The Pareto-superiority of E_U is illustrated in Table 1 (remember that at the equilibrium $g^N = g^{NB}$).

The situation captured in Fig. 2 (an unstable equilibrium surrounded by two stable ones) implies the existence of a threshold such that it is optimal for R to follow a policy leading in the long-run towards the stable equilibrium $E_L = (0, \lambda_L)$, whenever the initial value π_0 of π is inferior to the threshold value, $\pi_0 < \pi_s$ in other words, when the initial fraction of Bs is less than π_s, it is rational for the regulator to let the number of believers go to zero, although the resulting stationary equilibrium is associated with poor outcomes for all agents, R, Bs, and NBs.

Whenever $\pi_0 < \pi_s$ on the other hand, the regulator's optimal policy leads to the stable equilibrium E_U, that is, to a situation where there is in permanence, a positive fraction of believers.

The existence of the threshold is closely associated with the properties of the π-dynamics Eq. (6). For any given actions t^a, t (that is equivalently: for any given value of $g^B - g^{NB}$), the value of $\dot{\pi}$ will be smaller for small or large values of π than for intermediate ones. Note that, when a threshold exists and for any $\pi_0 < 1$, it is never optimal for the regulator to follow a policy that would ultimately insure that all firms are believers, $\pi \rightarrow 1$. There are two concurrent reasons for this: Once the π increases, the regulator has to deviate more and more from $t^{\$}(\pi)$ in order to make believing more profitable than not believing. On the other hand, $\dot{\pi}$ decreases. Thus, the discounted benefits of increasing π decrease. The rationale for steering the inferior equilibrium E_L starting from any $\pi_0 > \pi_s$ lies in the following fact: The "deviation costs" that the regulator will have to accept initially in order to steer the system towards the more favorable E_U are so high that the net benefit of going to E_U would be less than the one obtained by letting the system optimally converge towards the inferior E_L.

The above results suggest that the policy of an environmental regulator maybe very different depending on the fraction of the population that trusts it when it initiates a new policy. If the initial confidence is high, the costs of building a large fraction of *B*s are compensated in the long-run by accrued benefits. However, if the regulator's trustworthiness is low from the onset, its interest is to exploit its initial credibility for making short-term gains, although this ultimately leads to an inferior situation where nobody trusts the regulator.

As already mentioned, the dynamics of the canonical system Eqs. (6)–(23) at the unstable middle equilibrium E_M form a spiral. Therefore, the threshold value π_s is typically close to, but distinct from π_M (i.e., $\pi_M \neq \pi_s$) and the threshold takes the particularly challenging form of a Skiba point. No local analytical expression exists that characterizes a Skiba point. Thus, Skiba points must be computed numerically. This was not done in this paper. For a reference article on thresholds and Skiba points in economic models, see Deissenberg *et al.* (2004).

5. Sensitivity Analysis and Bifurcations Towards a Unique Equilibrium

Numerical analyses show that an increase in the firms' flexibility (i.e., an increase in β) shifts the stable equilibrium E_U to the right and the unstable equilibrium E_M as well as the Skiba point π_s to the left. In other words, if the firms react quickly to profit differences, the regulator will follow a policy that leads to the upper equilibrium E_U even if the initial proportion of *B*s is small. Indeed, the high speed of reaction of the firms means that the accumulated costs incurred by the regulator on the way to E_U will be small and easily compensated by the gains around and at E_U. Moreover, *R* does not have to make *B*s much better off than *NB*s to insure a fast reaction. As a result, for β large, the equilibrium E_U is characterized by a large proportion of *B*s and thus insures high stationary gains to all actors.

Not surprisingly, an increase of the discount factor ρ has the opposite effect. Impatient regulators want to build up confidence, only if the initial proportion of *B*s is high. The resulting equilibrium value π_U will be relatively low, since the time and efforts needed now for an additional increase in the stock of *B*s weighs heavily compared to the expected future benefits.

An increase of the prediction cost δ finally, shifts E_M to the left and E_U to the right: It is easier for *R* to incite the firms to use the *B* strategy, favoring the convergence to equilibrium with a large number of believers.

The scenario with two stable equilibria separated by a Skiba point can change qualitatively, when the parameter values are sufficiently altered. For small values of β and/or δ and/or for large values of ρ, the equilibrium E_U collides with E_M and disappears. Only one (stable) equilibrium remains, E_L. Thus, in the long-run, the proportion of believers tends to zero. On the other hand, for large values of δ trying to outguess the regulator is never optimal. The only remaining equilibrium is the one where everybody is a believer.

6. Conclusion

The starting point of this paper is a static situation, where standard rational behavior by all agents leads to a Pareto-inferior outcome, although there is no fundamental conflict of interest either among the different firms or between the firms and the regulator, and although the firms are perfectly informed and perfectly predict the regulator's actions. In addition to the environmental problem studied here, such a situation is common for instance in monetary economics, in disaster relief, patent protection, capital levies, or default on debt.

The present paper and previous work strongly suggest that, in such a situation, the existence of believers who take the announcements of a macroeconomic regulator at face value typically, Pareto-improve the economic outcome. This property hinges crucially on the fact that the private sector is atomistic, and thus, that the single agents do not anticipate the collative impact of their individual decisions. Moreover, it requires a relatively benevolent regulator, whose objective function does not differ too drastically from the one of the private agents.

To introduce dynamics, we assumed that the proportion of believers and non-believers in the economy changes over time as a function of the difference in profits for the two types of firms. The change obeys a word-of-mouth process driven by the relative success of the one or the other private strategy, to believe or to predict. It is supposed that the regulator recognizes its ability to influence the word-of-month dynamics, by an appropriate choice of actions and is interested not only in its instantaneous but also in its future losses. It is shown that, when the firms react *fast enough* to the difference in profits of *B*s and *NB*s, a *sufficiently patient* regulator may use incorrect announcements to Pareto-improve the economic outcome compared to the situation where all firms are non-believers that perfectly predict the regulator's actions. A particularly interesting case arise, when a stable Pareto-superior equilibrium with a positive fraction of believers co-exists with the (likewise stable) perfect prediction but inferior equilibrium where all firms are non-believers.

In this case, the optimal policy of the regulator (to converge towards the one or the other equilibrium) depends upon the initial proportion of believers in the population. This history dependence of the solution suggests a promising meta-analysis of the importance of good reputation based on the previous good governance, in situations where the same macro-economic policy-maker is confronted with recurrent, distinct decision-making issues.

These results of this paper are obtained under the assumption that the non-believers do not exploit the fact that the regulator announced tax differs from the equilibrium announcement for the static game. Likewise, in the model, the non-believers do not attempt to predict the dynamics of the number of believers. We leave to future research, the non-trivial task of exploring the consequences of a relaxation, of these two restrictive hypotheses.

References

Abrego, L and C Perroni (1999). Investment subsidies and time-consistent environmental policy. *Discussion Paper*, Coventry: University of Warwick, 1–35.

Barro, R and D Gordon (1983). Rules, discretion and reputation in a model of monetary policy. *Journal of Monetary Economics* **12**, 101–122.

Batabyal, A (1996a). Consistency and optimality in a dynamic game of pollution control I: competition. *Environmental and Resource Economics* **8**, 205–220.

Batabyal, A (1996b). Consistency and optimality in a dynamic game of pollution control II: monopoly. *Environmental and Resource Economics* **8**, 315–330.

Dawid, H (1999). On the dynamics of word of mouth learning with and without anticipations. *Annals of Operations Research* **89**, 273–295.

Dawid, H and C Deissenberg (2005). On the efficiency-effects of private (dis)-trust in the government. *Journal of Economic Behavior and Organization* **57**, 530–550.

Dawid, H, C Deissenberg and P Ševčík (2005). Cheap talk, gullibility and welfare in an environmental taxation game. In: *Dynamic Games: Theory and Applications*, A Haurie and G Zaccour (eds.), Amsterdam: Kluwer, 175–192.

Deissenberg, C and F Alvarez Gonzalez (2002). Pareto-improving cheating in a monetary policy game. *Journal of Economic Dynamics and Control* **26**, 1457–1479.

Deissenberg, C, G Feichtinger, W Semmler and F Wirl (2004). Multiple equilibria, history dependence, and global dynamics in intertemporal optimization models. In *Economic Complexity: Non-Linear Dynamics, Multi-agents Economies, and Learning*, W Barnett, C Deissenberg and G Feichtinger (eds.), ISETE, Vol. 14, Amsterdam: Elsevier, 91–122.

Dijkstra, B (2002). Time consistency and investment incentives in environmental policy. *Discussion Paper 02/12*.U.K.: School of Economics, University of Nottingham, 2–12.

Gersbach, H and A Glazer (1999). Markets and regulatory hold-up problems. *Journal of Environmental Economics and Management* **37**, 151–164.

Hofbauer, J and K Sigmund (1998). *Evolutionary Games and Population Dynamics*. Cambridge: Cambridge University Press.

Kydland, F and E Prescott (1977). Rules rather than discretion: the inconsistency of optimal plans. *Journal of Political Economy* **85**, 473–491.

Marsiliani, L and T Renstrom (2000). Time inconsistency in environmental policy: tax earmarking as a commitment solution. *Economic Journal* **110**, C123–C138.

McCallum, B (1997). Critical issues concerning central bank independence. *Journal of Monetary Economics* **39**, 99–112.

Petrakis, E and A Xepapadeas (2003). Location decisions of a polluting firm and the time consistency of environmental policy. *Resource and Energy Economics* **25**, 197–214.

Simaan, M and JB Cruz (1973a). On the Stackelberg strategy in nonzero-sum games. *Journal of Optimization Theory and Applications* **11**, 533–555.

Simaan, M and JB Cruz (1973b). Additional aspects of the Stackelberg strategy in nonzero-sum games. *Journal of Optimization Theory and Applications* **11**, 613–626.

Vallee, T (1998). Comparison of different Stackelberg solutions in a deterministic dynamic pollution control problem. *Discussion Paper, LEN-C3E*. France: University of Nantes, 1–30.

Vallee, T, C Deissenberg and T Basar (1999). Optimal open loop cheating in dynamic reversed linear-quadratic Stackelberg games. *Annals of Operations Research* **88**, 247–266.

Chapter 4

Modeling Business Organization

Topic 1

Enterprise Modeling and Integration: Review and New Directions

Nikitas Spiros Koutsoukis

Hellenic Open University, and Democritus University of Thrace, Greece

1. Introduction

Enterprise modeling (EM), Enterprise integration (EI), and Enterprise integration modeling (EIM) are concerned with the definition, analysis, re-design and integration of information, human capital, business process, processing of data and knowledge, software applications and information systems regarding the enterprise and its external environment to achieve advancements in organization performance (Soon *et al.*, 1997).

Enterprise modeling and Enterprise integration still are still ongoing fields of research where the background and interests of contributors in these areas are heterogeneous, including disciplines like mechanical manufacturing, business process re-engineering, and software engineering.

However, the following two definitions (Vernadat, 1996) are representative for current practices:

- Enterprise modeling is concerned with the explicit representation of knowledge regarding the enterprise in terms models.
- Enterprise integration consists of breaking down organizational barriers to improve synergy within the enterprise so that business goals are achieved in a more productive way.

The main objective of this paper is to review the literature on EM and integration and to provide a concise description of related topics. The paper also discusses EM and integration challenges, and rationale as well as current tools and methodologies for deeper analysis.

2. Model and Enterprise Terminology

Model and enterprise are the two key terms that appear throughout this paper. An enterprise can be perceived as "a set of concurrent processes executed on the enterprise means (resources or functional entities) according to the enterprise objectives and subject to business or external constraints" (Vernadat, 1996). A model can be defined as "a structure that a system can use to simulate or anticipate the behaviour of something else" (Hoyte, 1992). This is a generalization of a definition given by Minsky (1968): "a model is a useful representation of some subject. It is a (more or less formal) abstraction of a reality (or universe of discourse) expressed in terms of some formalism (or language) defined by modeling constructs for the purpose of the user".

Instead of attempting to form a definition, which may be redundant, we will mention some important features of enterprise models instead.

- An enterprise model provides a computational representation of the structure, activities, processes, information, resources, people, behavior, goals, and constraints of a business, government, or other enterprise. It can be both descriptive and definitional-spanning what is and what should be. The role of an enterprise model is to achieve model-driven enterprise design, analysis, and operation (Fox, 1993).
- Enterprise models must either be represented in the mind of human beings or in computers (Christensen *et al.*, 1995).
- All enterprise models are built with a particular purpose in mind. Depending on the purpose, the model will emphasize different types of information at different levels of detail.

3. Enterprise Modeling

During the last decades, EM has been partly addressed by several approaches to the same problem of making the enterprise more efficient. All these approaches have included the construction of symbolic representations of aspects of the real world, using various notations and techniques that commonly could be denoted "conceptual modeling" (Solvberg and Kung, 1993). The idea of EM, as an extension of the standard for the exchange of product model data (STEP) product modeling approach, contributes to integrating design (which focuses on product information) and construction (focusing on activity and resource information) (Kemmerer, 1999).

In the 1980s and during the early 1990s, when database technology was applied on a larger scale, the need for better application design and development methods became evident (Martin, 1982; Sager, 1988). Different modeling methodologies, like the structured analysis and design technique (SADT), were developed. A narrow focus on a single department or task during development often caused operational problems across application boundaries. The need for making more complete models aroused, integrating operations across departmental or functional boundaries.

The problem in EM regarding the manufacturing of EI and control was to ensure timely execution of business processes on the functional entities of the enterprise (i.e., human and technical agents) to process enterprise objects. Processes are made of functional operations and functional operations are executed by functional entities. Objects flowing through the enterprise can be information entities (data) as well as material objects (parts, products, tools, etc.) (Rumbaugh, 1993).

In order to design the flexible information and communication infrastructure needed, different enterprise models are useful, and the model may in fact become a part of the integrated enterprise itself. The trend now is away from the first single perspective enterprise models, which was only focusing on the product data information handling, to more complete enterprise models covering both informational and human aspects of the organization (Fox, 1993).

During the 1990s, the question aroused on how to integrate different departments in an enterprise, and how to connect the enterprise with its external environment i.e., suppliers and customers, to improve co-operation and logistics. During the 1990s, conglomerate industries took a more developmental approach, and the research area of EI emerged. Enterprise modeling is clearly a pre-requisite for EI as all things to be integrated and coordinated need to modeled to some extent (Petrie, 1992).

Since 1990, major R&D programs for computer-integrated manufacturing (CIM) have resulted in the following tools/methods or architectures for EM:

(1) IDEF modeling tools: These are complementary modeling tools developed by the integrated computer aided manufacturing (ICAM) project, introducing IDEF0 for functional modeling, IDEF1 for information modeling, IDEF2 for simulation modeling, IDEF3 for business process modeling, IDEF4 for object modeling, and IDEF5 for ontology modeling. IDEF models are mainly used for requirements definitions.
(2) Generic Artificial Intelligence (GRAI) integrated methodology: It is a methodology which uses the GRAI grid, GRAI nets, developed by the

GRAI laboratories of France and includes modeling IDEF0, and group technology to model the decision making processes of the enterprise. The combination is called GRAI integrated methodology (GIM).

(3) Computer integrated manufacturing open system architecture (CIM-OSA): CIMOSA has been developed to assist enterprises adapt to internal and external environmental changes so as to face global competition and improve themselves regarding product prices, product quality, and product delivery time. It provides four different types of views of the enterprise: modeling function view, information view, resource view, and organization view (Beeckman 1993; Vernadat, 1996).

(4) Purdue enterprise reference architecture (PERA): It is a methodology developed at Purdue University in 1989 which focuses on separating the human-based functions of an enterprise from the manufacturing and information functions (Williams, 1994).

(5) Generalized enterprise reference architecture and methodology (GERAM): It is a generalization of CIMOSA, GIM, and PERA. The general concepts, identified and defined in this reference methodology, consist of methodological guidelines for enterprise engineering (from PERA and GIM), life-cycle guidelines (from PERA), and model views modeling (e.g., CIMOSA constructs).

Modeling method relies upon a specific purpose defining its finality, i.e., the goal of the modeler. This finality has a direct influence on the definition of modeling method. We adopt the position that any enterprise is made of a large collection of concurrent business processes, executed by an open set of functional entities (or agents) to achieve business objectives (as set by management). Enterprise modeling and integration is essentially a matter of modeling and integrating these process and agents (Kosanke and Nell, 1997; Ladet and Vernadat, 1995).

The prime goal of an EM approach is to support analysis of an enterprise rather than to model the entire enterprise, even though this is theoretically possible. In addition, another goal is to model relevant business process and enterprise objects concerned with business integration.

According to Vernadat (1996), the aim of EM is to provide:

- a better perspective of the enterprise structure and operations;
- reference methods for enterprise engineering of existing or new parts of the enterprise both in terms of analysis, simulation, and decision-making and
- a model in order to manage efficiently the enterprise operations.

Whereas, the main motivations for EM are:

- better understanding of the structure and functions of the enterprise;
- managing complexity of operations;
- creating enterprise knowledge and know-how for all;
- efficient management of processes;
- enterprise engineering and
- EI itself.

4. Enterprise Integration

Enterprise integration was discussed since the early days of computers in industry and especially in the manufacturing industry with CIM as the acronym for operations integration (Vernadat, 1996). In spite of the different understandings of the scope of integration in CIM, it was always understood for information integration across or at least parts of the enterprise. Information integration essentially consists of providing the right information, to the right people, at the right place, at the right time (Norrie *et al.*, 1995). Information that is available to the right people, at the right place, at the right time, and that enables them to act quickly and decisively, is a crucial requirement of enterprises that intend to grow and remain profitable. The intelligent delivery of information must obtain maturity levels consistent with enterprise goals and must employ the appropriate technology to facilitate this purpose. In today's enterprise environment, users must consolidate data and deliver it as useful information and knowledge that empowers the right business decisions.

With this understanding the different needs in EI can be identified:

(1) Identify the right information to gain knowledge: Elicit knowledge from gathering information from the intra-enterprise activities and use models to represent this knowledge regarding product attributes and information, process management issues, and the decision-making process.
(2) Provide the right information at the right place: The need for implementing information sharing systems and integration platforms to support information transaction across heterogeneous environments regarding hardware, different operating systems, and the legacy systems. This is a step towards inter-enterprise integration overcoming barriers and provides connection between enterprise environments by linking their operations to create extended and virtual enterprises.

(3) Up-date the information in real time to reflect the actual state of the enterprise operation: enterprises must be able to adopt changes regarding both their external environment i.e., technology, legislation, and their internal environment i.e., production operations in order to protect the enterprise goal and objective from becoming obsolete.
(4) Co-ordinate business processes: This requires decisional capabilities and know-how within the enterprise for real-time decision support and evaluation of operational alternatives based on business process modeling.

The aims of EI are:

- to enable communication among the various functional entities of the enterprise;
- to provide interoperability of IT applications (plug-and-play approach) and
- to facilitate coordination of functional entities for executing business processes so that they synergistically fulfill enterprise goals.

The main motivation for EI is to make possible true interoperability among heterogeneous, remotely located, systems within, or outside the enterprise in an independent IT vendor environment.

Enterprise integration, especially for networked companies and large manufacturing enterprises is becoming a reality. The issue is that EI is both an organizational and a technological matter. The organizational matter is partly understood so far whereas the technological matter lies with the major advances over the last decade. It concerns mostly computer communications technology, data exchange formats, distributed databases, object technology, Internet, object request brokers (ORB such as OMG/CORBA), distributed computing environments (such as OSF/DCE and MS DCOM, and now Java to enterprise edition and execution environments (J2EE).

Some important projects developing integrating infrastructure (IIS) technology for manufacturing environments include:

- CIMOSA IIS: The CIMOSA integrated infrastructure (IIS) enables business process control in a heterogeneous IT environment. In IIS, an enterprise is composed of functional entities, which are connected by a communication system. Functional entities can be a machine, a software application, or a human (Vernadat, 2002).
- AIT IP: This is an integration platform developed as an AIT project and based on the CIMOSA IIS concepts (Waite, 1998).

- OPAL: This is also an AIT project that has proved that EI can be achieved in design and manufacturing environments using existing ICT solutions (Bueno, 1998).
- NIIIP (National industrial information infrastructure protocols) (Bolton *et al.*, 1997).

However, EI suffers from inherent complexity of the field, lack of established standards, which sometimes happen after the fact, and the rapid and unstable growth of the basic technology with a lack of commonly supported global strategy. The EI modeling research community is growing at a very rapid pace (Fox and Gruninger, 2003). The EIM research so far has been mostly performed in large research organizations. A few pilot studies have also been done in large organizations. Hunt (1991) suggests that EIM is most beneficial to large industrial and government organizations. The emphasis of current EIM research on large organizations is risky, and may even have hampered the growth of the field. Enterprise wide modeling is difficult in large organizations for two reasons: there is seldom any consensus on exactly how the organization functions and the modeling task may be so big that by the time any realistic model is built, the underlying domain may have changed. Whatever these terms mean, we have to seek solutions to solve industry problems, and this applies to EM and integration.

5. A Framework Setting for EI Modeling

Enterprise integration is a multifaceted task employed for addressing challenges stemming from various aspects of the globalized economy. The contemporary business world, economically, politically, and technologically is a fast-changing world and most if not all enterprises, from small to large, are continually challenged to evolve as rapidly as the world around them evolves. Enterprise integration allows an enterprise to function coherently as a whole and to shift its primary value chain activities in tandem with its secondary value chain activities according to the challenges or requirements posed by the enterprise environment or goals set by the enterprise itself, and most frequently a combination of both (Giachetti, 2004). In this context, the value chain's primary and secondary activities are used simply as a well-known functional-economic view of the enterprise in order to communicate coherently, as opposed to a semantically accurate or complete view of the enterprise (Porter, 1985).

The multiple facets of EI refer to the multiple dimensions, entities, and constituents where integration is primarily aimed at, and through which enterprise integration is achieved.

Enterprise integration dimensions represent infrastructural views, upon which integration is founded, built and subsequently achieved. Some of the most commonly used integration dimensions are the following:

- the data and information dimension;
- the processes dimension;
- the economic activity or business dimension and
- the technology or technical dimension.

Because these dimensions can easily be extended throughout the enterprise regardless of the enterprise functions, they are frequently used as fundamental reference points for EI efforts (Lankhorst, 2005).

Each of these dimensions has been used successfully for seeking out and achieving EI. For instance, in the data and information dimension, which is mainly the realm of corporate information systems, data warehousing and enterprise resource planning (ERP), have been used successfully to introduce enterprise-wide, unified views of data, information and related analyses. In the processes dimension, business process management (BPM) or re-engineering (BPR) have been successfully applied in developing and refining old, and new processes which are more integrated and efficient across the enterprise. In the economic or business dimension, business strategy models, like the core-competency theory, balanced-score cards, or economic value-added models like the activity-based costing (ABC) are also used successfully to integrate across the enterprise (Kaplan and Norton, 1992; Prahalad and Hamel, 1990). In the technology or technical dimension, the key requirement is for interoperability of systems, hardware, and software. The success with which many multinational enterprises operate shows that certain milestones have been reached in this direction.

Enterprise entities refer to function-oriented enterprise constructs that are typically goal-oriented, collectively contributing to achieving the enterprise mission. Entities form hierarchical views of the enterprise, and frequently tend to match formal or informal structures as in most organizational charts. From a functional perspective, entities refer to key organizational activities or divisions, such as administration, production, accounting, sales, marketing, procurement, and R&D. As is well known, entities can also be organized in terms of authority or delegation layers, communication layers, goal-contribution layers, geographically based activities and so forth.

The sole purpose for having such functional-oriented constructs is to decompose the enterprise into smaller, more specific, and more manageable parts. The contribution of each entity to the enterprise objectives is thus readily identified and narrowed down in scope, horizontally or vertically. It is only through the use of one or more dimensions that the entities are re-combined in an enterprise-wide integration model.

Enterprise constituents or constituent groups refer to the human factor which is ultimately responsible for setting out, applying and achieving EI. The most prevalent constituent groups are the following:

- Stakeholders
- Decision-makers
- Employees

which are internal to the enterprise, and

- Suppliers and
- Clients

which are external to the enterprise.

Some of the main pitfalls in EI lie at the constituents' level. This is because the constituents are, in one way or another, autonomous or semi-autonomous agents acting primarily in their own interest, which is defined mainly by their position in enterprise coordinates terms. We consider the following fundamental dimensions as the main enterprise coordinates for positioning the constituents:

- level of authority and
- immediacy of contribution to goals (from strategic to operational, abstract to specific)

which can also be inverted to

- level of responsibility and
- level of activity (from strategic to operational, or abstract to specific)

The higher the level of authority, the more strategic or abstract are the actions that the constituents take in enterprise terms. At this level, constituents have a bird's-eye view of the enterprise and find it easier to match organizational outcomes to personal interest i.e., the bird's-eye view makes it easier to view the enterprise as a single entity. On the other hand, the lower the level of authority the more operational or specific the actions that the constituents take in enterprise terms. At this level, constituents find it harder to consider a single-entity view of the organization, and hence organizational outcomes are less easily matched to personal interests.

It follows that inconsistencies in the constituents' perspectives have the ability to cancel out some of the main benefits that the other two facets bring forward, namely the unified view of the enterprise as single entity and the immediacy with which operational activities contribute to strategic goals. Inconsistencies in constituents' perspectives are being addressed mainly through leadership and human resource practices, which aim to put in place an organizational culture, or otherwise a unified perception of what the enterprise stands for, its values and its practices among other things. Mission statements, codes of practice, and team building exercises are some of the most frequently used tools for scoping and setting out a unified, constituent view of the enterprise.

Thus, a strong requirement is surfacing for developing new EI modeling frameworks, which are able to incorporate heterogeneous modeling constructs, such as the dimensions, entities or constituents considered above. We find that some of the key requirements are set out as follows:

(a) Interchangeable dimensions: EI models should be able to shift between the dimensions considered above, without loss of context.
(b) Use of hybrid modeling constructs: EI models should be able to utilize interchangeably modeling constructs for representing any type of enterprise activity unit, including procedural units, physical or logical units, inputs or outputs, tangible and intangible processes, decision-making points, etc.
(c) Ability to consolidate or analyze to successive levels of detail: this is particularly useful for communicating enterprise structures throughout multiple echelons and hierarchy levels. For instance, stakeholders will most likely need access to a consolidated view of the integrated enterprise whereas employees are most likely to require a position-centred view of the enterprise, with detailed surroundings that are consolidated further away from the employee's position.
(d) Ability to interchange between declarative and goal-seeking states of the integration models: when in declarative mode, models work simply as formal structures, or simulators of how the integrated enterprise operates. In goal-seeking mode, the model emphasizes improvements that can lead to goal attainment or "satisfying," to use Simon's term (Simon, 1957).

Of course, these requirements can be analyzed further and in detail, but such a discussion would be beyond the scope of this paper. Still, we also find that these requirements are suffice to set out an agenda for further work in the domain of EI and EI modeling.

6. Discussion

At the current state of technology, we find that EM and integration are becoming familiar within the large enterprises that seem to understand the need of acquiring their concept. However, in accordance with Williams (1994) we also find that the reason for the moderate success and rather not so extended use of EM and integration initiatives include:

(a) Cost: It is commonly accepted that such efforts are very expensive and it is rather difficult to argue on the benefits of the outcome from the beginning of each project.
(b) Project size and duration: Indeed, such projects cannot be completed but they require time. Quite a large number of people will be involved and high levels of commitment on behalf of the enterprise are required.
(c) Complexity: EI is a never-ending process and parameters such as the legacy systems within large enterprises resisting integration have to be seriously considered. There is the need for a global vision with a modular implementation approach.
(d) Management support: Management facing high cost may resent to support such projects. It is very important that management be part of the project and be fully committed to it.
(e) Skilled people: The human factor concerning high-skilled employees involved in the projects is vital for success. Unfortunately, due to limited training schemes experienced users are scarce.

Nevertheless, the enthusiasm of the research community and the industry remains intact trying to overcome problems and difficulties by creating tools and adopting methodology that will make a difference. Nowadays, EM and integration mostly rely upon advanced computer networks, the worldwide web and Internet, integration platforms as well as ERPs, and data exchange formats (Chen and Vernadat, 2002; Martin *et al.*, 2004).

A big issue is that both EM and EI are nearly totally ignored by SME's and there is still a long way to go before SME's will master these techniques on their own and in this case, they have to quickly catch on the technology and define which tools they have to use.

Some researchers like Vernadat, Kosanke, Tolone, and Zeigler consider that this technology would better penetrate and serve any kind of enterprises if:

- They identify users and a specific EIM problem. An example of an EIM problem is how to produce high-quality products at low cost.
- They make a prototype EIM that addresses the problem, contributing in the decision-making process based on a simple initial model accepted by the majority of the users (Tolone *et al.*, 2002).
- There was a standard vision on what EM really depend and there was an international consensus on the underlying concept for the benefit of the business users (Kosanke *et al.*, 2000).
- There was a standard, user-oriented, interface in the form of a unified enterprise modeling language (UEML) based on the previous consensus to be available on all commercial modeling tools.
- There were real EM and simulation tools commercially available in the market place taking into account function, information, resource, organization, and financial aspects of an enterprise including human aspects, exception handling, and process coordination. Simulation tools need to be configurable, distributed, and agent-based simulation tools (Vernadat and Zeigler, 2000).

However, it is rather unlikely that a single EIM that will be equally applicable to all organizations. Each organization will have to develop its own strategy and philosophy (Fox and Huang, 2005). Despite the different approaches (frameworks) of EIM, we always have to bear in mind the following functional requirements of an EIM framework (Patankar and Adiga, 1995):

(1) Decision orientation. An EI model based on any particular concept of an EM framework should support the decision-making process considering thoroughly all factors of the external and internal environment affecting the enterprise.
(2) Completeness. EM frameworks should be able to represent all aspects of the enterprise such as business units, business data, information systems, and external partners i.e., suppliers, sub-contractors, and customers.
(3) Appropriateness. EIM frameworks should be representing the enterprise at appropriate levels by providing models that can be understood by people with heterogeneous backgrounds.

(4) Permanence. The EI model should remain consistent with the enterprise's overall objectives even when the enterprise is in the process of changing due to management changes or other. Models should remain true and valid within the enterprise.

Finally, the EIM frameworks should be simply structured, flexible, modular, and generic to ensure that the framework is applicable across different manufacturing operations and enterprises. In addition, the framework should relate to both manual and automatic processes. This will provide the basis for proper and complete integration.

Modeling and integration always mean effort in time and costs. Nevertheless, to handle complexity in business environments we see the above-mentioned elements moving from "luxury articles" to "everyday necessities".

These are some of the challenges to be solved in future to build extended interoperable manufacturing enterprises as there is unlikely to be a single EIM methodology, which will be equally applicable to all organizations (Li and Williams, 2004). Each organization will have to develop its own strategy and philosophy. However, EIM promises to be a revolutionary technology.

Finally, the success of EI efforts lies in the ability to iron out incoherencies and lack coordination between the multiple facets considered above. This is a complex and challenging task, and lies at the heart for EI modeling. Many of the multiple modeling frameworks considered above, like IDEF, CIM-OSA, or GERAM for example, have stemmed from integration efforts in the areas of manufacturing or engineering, and while successful in their own right, these frameworks are still lagging behind when it comes to integrating across the multiple dimensions, entities and constituents of a contemporary enterprise that we have identified above.

7. Epilogue

This paper has provided a description of the main issues on EM and EI. It has become quite evident from the literature that both the issues are complex and require knowledge on various disciplines particularly on organizational management, engineering, and computer science. As such, it can be generalized that EM and integration mainly revolves around organizational modeling from which the analysis, design, implementation and integration of the organization management, business processes, software applications, and physical computer systems within an enterprise are supported.

We believe that although we are rich in technology to solve isolated business problems (e.g., product scheduling, quality control, production planning, accounting, marketing, etc.), many other problems still reside within the enterprise. Solving these problems require an integrated approach to problem solving. This integration must start at the representation level and the entire enterprise together with its structure, functionality, goals, objectives, constraints, people, processes, products must be modeled. The solution will probably be the outcome of an EIM framework.

References

Beeckman, D (1993). CIM-OSA: computer integrated manufacturing open system architecture. *International Journal of Computer Integrated Manufacturing* **2**(2), 94–105.

Bolton, R, A Dewey, A Goldschmidt and P Horstmann (1997). NIIIP — the national industrial information infrastructure protocols. In *Enterprise Engineering and Integration: Building International Consensus*, K Kosanke and JG Nell (eds.), Berlin: Springer-Verlag. pp. 293–306.

Bueno, R (1998). Integrated information and process management in manufacturing engineering. *Proc. of the 9th IFAC Symposium on Information Control in Manufacturing (INCOM'98)*, pp. 109–112. France: Nancy-Metz, May 24–26.

Chen, D and F Vernadat (2002). *Enterprise interoperability: a standardisation view*. ICEIMT 2002, 273–282.

Christensen, LC, BW Johansen, N Midjo, J Onarheim, TG Syvertsen and T Totland (1995). Enterprise Modeling-Practices and Perspectives. In *Proc. of 9th ASME Engineering Database Symposium*, Rangan (ed.), Boston: Massachusetts, 1171–1181.

Fox, MS (1993). Issues in Enterprise Modeling. *Proc. of the IEEE Conference on Systems*. Man and Cybernetics, France: Le Touquet, 83–98.

Fox, MS and M Gruninger (2003). The logic of enterprise modeling, modeling and methodologies for enterprise integration. In *Modelling and Methodologies of Enterprise Integration*, P Bernus and L Nemes (eds.), Cornwall, Great Britain: Chapman and Hall, 83–98.

Fox, MS and J Huang (2005). Knowledge provenance in enterprise information. *International Journal of Production Research* **43**(20), 4471–4492.

Giachetti, RE (2004). A framework to review the information integration of the enterprise. *International Journal of Production Research* **42**(6), 1147–1166.

Hoyte, J (1992). Digital models in engineering — a study on why and how engineers build and operate digital models for decision support. PHD Thesis, University of Trondheim, Norway.

Hunt, DV (1991). *The Integration of CALS, CE, TQM, PDES, RAMP, and CIM, Enterprise Integration Sourcebook*. San Diego, CA: Academic Press.

Kaplan, RS and DP Norton (1992). The balanced scorecard: measures that drive performance. *Harvard Business Review* **70**(1), 71–79.

Kemmerer, J (1999). *STEP: The Grand Experience*, NIST Special Publication SP939, Washington, C 20402, USA: US Government Printing Office.

Kosanke, K and JG Nell (1997). *Enterprise Engineering and Integration: Building International Consensus*. Berlin: Springer-Verlag.

Kosanke, K, F Vernadat and M Zelm (2000). Enterprise engineering and integration in the global environment. *Advanced Network Enterprises 2000*, Berlin, 61–70.
Ladet, P and FB Vernadat (1995). *Integrated Manufacturing Systems Engineering*. London: Chapman & Hall.
Lankhorst, M (2005). *Enterprise Architecture at Work*. Berlin: Springer-Verlag.
Li, H and TJ Williams (2004). *A Vision of Enterprise Integration Considerations*. Toronto, Canada: ICEIMT 2004.
Martin, J (1982). *Strategic Data-Planning Methodologies*. Englewood Cliffs, New York: Prentice-Hall.
Martin, J, E Robertson and J Springer (2004). *Architectural Principles for Enterprise Frameworks: Guidance for Interoperability*. Toronto, Canada: ICEIMT 2004.
Minsky, M (1968). *Semantic Information Processing*. Cambridge, MA: The MIT Press.
Norrie, M, M Wunderli, R Montau, U Leonhardt, W Schaad and HJ Schek (1995). Coordination approaches for CIM. In *Integrated Manufacturing Systems Engineering*. P Ladet and FB Vernadat (eds.), London: Chapman & Hall pp. 221–235.
Patankar, AK, and S Adiga (1995). Enterprise integration modeling: a review of theory and practice. *Computer Integrated manufacturing Systems* **8**(1), 21–34.
Petrie, CJ (1992). *Enterprise Integration Modeling*. Cambridge, MA: The MIT Press.
Porter, M (1985). *Competitive Advantage*. New York: Free Press.
Prahalad, CK and G Hamel (1990). The core competence of the organisation. *Harvard Business Review* **68**(3), 79–91.
Rumbaugh, J (1993). Objects in the constitution — enterprise modeling. *Journal on Object Oriented Programming*, January issue, 18–24.
Sager, MT (1988). Data concerned enterprise modeling methodologies — a study of practice and potential. *The Australian Computer Journal* **20**(3), 59–67.
Simon, H (1957). *Models of Man: Social and Rational*. New York: Wiley.
Solvberg, A and DC Kung (1993). Information Systems Engineering — An Introduction. New York: Springer-Verlag.
Soon, HL, J Neal and A Pennington (1997). Enterprise modeling and integration: taxonomy of seven key aspects. *Computers in Industry* **34**, 339–359.
Tolone, W, B Chu, G.-J Ahn, R Wilhelm and JE Sims (2002). Challenges to Multi-Enterprise Integration. *ICEIMT 2002*, 205–216.
Vernadat, FB (1996). *Enterprise Modeling and Integration: Principles and Applications*. London: Chapman & Hall.
Vernadat, F (2002). Enterprise modeling and integration. *ICEIMT 2002*, 25–33.
Vernadat, FB and BP Zeigler (2000). New simulation requirements for model-based Enterprise Engineering. *Proc. of IFAC/IEEE/INRIA Internal Conference on Manufacturing Control and Production Logistics (MCPL'2000)*. Grenoble, 4–7 July 2000. CD-ROM.
Waite, EJ (1998). Advanced Information Technology for Design and Manufacture. In *Enterprise Engineering and Integration: Building International Consensus*. K Kosanke and JG Nell (eds.), Berlin: Springer-Verlag, pp. 256–264.
Williams, TJ (1994). The purdue enterprise reference architecture. *Computers in Industry*, **24**(2–3), 141–158.

Topic 2

Toward a Theory of Japanese Organizational Culture and Corporate Performance

Victoria Miroshnik
University of Glasgow, UK

1. Introduction

Japanese national culture has a specific influence on the organizational culture (OC) of the Japanese companies. In order to understand these aspects of culture that are affecting corporate performance, a systematic analysis of the national culture (NC) and OC is essential. The purpose of this chapter is to provide a scheme to analyze this issue in a systematic way.

National culture can affect corporate OC (Aoki, 1990; Axel, 1995; Hofstede, 1980) and corporate performance (CP) (Cameron and Quinn, 1999; Kotter and Heskett, 1992). Multinational corporations with its supposedly common OC demonstrate different behaviors in different countries, which can affect their performances (Barlett and Yoshihara, 1988; Calori and Sarnin, 1991). National culture with its various manifestations can create unique organizational values for a country, which in conjunction with the human resources management (HRM) practice may create some unique OC for a corporation. Organizational culture along with a leadership style, which is the product of OC, NC, and HRM practices, creates a synergy, which shapes the fortune of the company's performances. Thus, NC, organizational values, OCs, leadership styles, human resources practices, and CPs are inter-related and each depend on the other. If a company maintains this harmony it can perform well, otherwise it declines.

The structure of this chapter is as follows. In Section 2, relationship between NCs, OC, and performances, as given in the existing literature, is analyzed. In Section 3, the Japanese culture was analyzed along with the controversies on this issue. In Section 4, the new approach of the nation-organization-performance exchange model (NOPX) is explained in detail and the theoretical propositions underlying this model are analyzed.

2. Corporate Performance, the Highest Stage of National and Organizational Culture: A Synthesis

Culture is multidimensional, comprising of several layers of inter-related variables. As society and organizations are continuously evolving, there is no theory of culture valid at all times and locations. The core values of a society (*macro-values* (MAVs)) can be analyzed by asking the following questions:

(a) What are the relationships between human and nature? (b) What are the characteristics of innate human nature? (c) What is the focus regarding time — whether past, future, or present, or a combination of all these? (d) What are the modalities of human activity, whether spontaneous or introspective or result-oriented or a combination of these? (e) What is the basis of relationship between one man and another in the society?

In the structure of the NC, there are certain *micro or individual values*, which include sense of belonging, excitement, fun and enjoyment, warm relationship with others, self-fulfilment, being well-respected, and a sense of accomplishment, security, and self-respect (Kahle, *et al.*, 1988). Although Hofstede (1990; 1993) has the opinion that the perceived practices are the roots of the OC rather than values, most researchers have accepted the importance of values in shaping cultures in several organizations.

Combinations of micro-values (MIVs) and macro-values (MAVs) can give rise to an OC, given certain meso-values (MEV) or *organizational values*, which are specific for a country. Meso-values are certain codes of behavior, which influences the future, and the expected behavior of the members of the society, if they want to belong to the main stream. These vary from one society to another, as the expectations of different societies, as products of historical experiences, religious, and moral values are different.

Cultural values are shared ideas about what is good, desirable, and justified; these are expressed in a variety of ways. Social order, respect for traditions, security, and wisdom are important for the society, which is like an extended family.

According to Schein (1997), OC emerges from some common assumptions about the organization, which the members share as a result of their experiences in that organization. These are reflected in their pattern of behaviors, expressed values, and observed artefacts. Organizational culture provides accepted solutions to known problems, which the members learn and feel about and forms a set of shared philosophies, expectations, norms, and behavior patterns, which promotes higher level of achievements (Kilman *et al.*, 1985; Marcoulides and Heck, 1993; Schein, 1997).

Organizational culture, for a large and geographically dispersed organization, may have many different cultures. According to Kotter and Heskett (1992), leaders create certain vision or philosophy and business strategy for the company. A corporate culture emerges that reflects the vision and strategy of the leaders and experiences they had while implementing these strategies. However, the question is whether it is valid for every NC.

A combination of MAVs, MEVs, and MIVs creates a specific OC, which varies from country to country according to their differences in NC. Hofstede (1980; 1985; 1990; 1993) showed the relationship between NC and OC. Global leadership and organizational behavior effectiveness (GLOBE) research project has tried to identify the relationships among leadership, social culture, and OC (House, 1999).

Cameron and Quinn (1999) have mentioned that the most important competitive advantage of a company is its OC. If an organization has a "strong culture" with "well-integrated and an effective" set of values, beliefs, and behaviors, it normally demonstrates high level of CPs (Ouchi, 1981).

Denison and Mishra (1995) have attempted to relate OC and performances based on different characteristics of the OC. However, different types of OC enhance different types of business (Kotter and Heskett, 1992). There is no single cultural formula for long-run effectiveness. As Siehl and Martin (1988) have observed, culture may serve as a filter for factors that influence the performance of an organization. These factors are different for different organizations. Thus, a thorough analysis regarding the relationship between culture and performances is essential.

3. Japanese National and Organizational Culture

Japanese firms are considered to be of a family unit with long-term orientations for HRM and with close ties with other like-minded firms, banks, and the government. The OC promotes loyalty, harmony, hard work, self-sacrifice, and consensus decision-making. These, along with lifetime employment and seniority-based promotions, are considered to be the natural outcome of the Japanese NC. (Ikeda, 1987; Imai, 1986; Ouchi, 1981). Japanese workers' psychological dependency on companies emerges from their intimate dependent relationships with the society and the nation.

Japanese NC has certain MIVs: demonstration of appropriate attitudes (*taido*); the way of thinking (*kangaekata*); and spirit (*ishiki*), which forms the basic value system for the Japanese (Kobayashi, 1980). Certain MEVs can be identified as the core of the cultural life for the Japanese, if they want to belong to the mainstream Japanese society (Basu, 1999; Fujino, 1998). These are (a) *Senpai-Kohai* system; (b) *Conformity*; (c) *Hou-Ren-Sou* and (d) *Kaizen* or continuous improvements.

Senpai-Kohai or senior-junior relationships are formed from the level of primary schools where junior students have followed the orders of the senior students, who in turn, may help the juniors in learning. The process continues throughout the lifetime for the Japanese. In work places, *Senpais* will explain *Kohais* how to do their work, the basic code of conducts and norms.

From this system emerges the second MEV, *Conformity*, which is better understood from the saying that *"nails that sticks out should be beaten down"*. The inner meaning is that unless someone conforms to the rules of the community or co-workers, he/she would be an outcaste. There is no room for individualism in the Japanese society or in work place.

The third item, *Hou-Ren-Sou*, is the basic feature of the Japanese organizations. *Hou-Ren-Sou*, is a combination of three different words in Japanese: *Houkoku* i.e., to report, *Renraku* i.e., to inform and *Soudan* i.e., to consult or pre-consult.

Subordinates should always report to the superior. Superiors and subordinates share information. Consultations and pre-consultations are required; no one can make his/her decision by himself/herself even within the delegated authority. There is no space in which the delegation of authority may function. Combining these words, *Houkoku*, *Renraku*, and *Soudan*, "*Hou-Ren-Sou*" is the core value of the culture in Japan. Making suggestions, for improvements, without pre-consultations is considered to be an offensive behavior in Japanese culture. Everyone from the clerk to the president, from the entry day of the working life to the date of retirement, every Japanese must follow the "*Hou-Ren-Sou*" value system.

Kaizen i.e., continuous improvement is another MEV that is one of the basic ingredients of the Japanese culture. Search for continuous improvement during the Meiji Government of the 19th century led them to search the world for knowledge. Japanese companies today are doing the same in terms of both acquisitions of knowledge by setting up R&D centers throughout the western world, and by having "*quality circles*" in work places in order to implement new knowledge to improve the product quality and to increase its efficiency (Kujawa, 1979; Kumagai, 1996; Miyajima, 1996).

3.1. *Japanese OC*

In order to understand the Japanese OC, it is essential to know the Japanese system of management, which gave rise to the unique Japanese style of OC. What is written below is the result of several visits to the major Japanese corporations (like Toyota, Mitsubishi, Honda, and Nissan) and interviews with several senior executives, including members of the board and vice-presidents of these companies.

Japanese system of management is a complete philosophy of organization, which can affect every part of the enterprise. There are three basic ingredients: lean production system, total quality management (TQM) and HRMs (Kobayashi, 1980). These three ingredients are inter-linked in order to produce total effect on the management of the Japanese enterprises. Because Japanese overseas affiliates are part of the family of the parent company, their management systems are part of the management strategy of the parent company (Basu, 1999; Morishima, 1996; Morita, 1992; Shimada, 1993).

The basic idea of the lean production system and the fundamental organizational principles is the "*Human-ware*" (Shimada, 1993), which is described in Fig. 1. "*Human-ware" is* defined as the integration and interdependence of machinery and human relations and a concept to differentiate among different types of production systems. The purpose of the lean production philosophy, which was developed at the Toyota Motor Company, is to lower the costs. This is done through the elimination of waste — everything that does not add value to the product. The constant strive for perfection (*Kaizen* in Japanese) is the overriding concept behind good management, in which the production system is being constantly improved; perfection is the only goal. Involving everyone in the work of improvement is often accomplished through quality circles. Lean production system uses "autonomous defect control" (*Pokayoke* in Japanese), which is an inexpensive means of conducting inspection for all units to ensure zero defects. Quality assurance is the responsibility of everyone. Manufacturing tasks are organized into teams. The principle of "*just-in-time*" (JIT) means, each

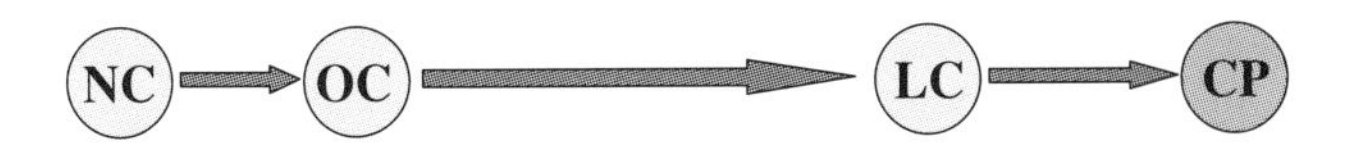

Where "NC" is "National Culture"; "OC" is "Organizational Culture"; "LC" is "Leader Culture"; and, "CP" is "Corporate Performance".

Figure 1. Sub-model 1: culture-performance system.

process should be provided with the right part, in the right quantity at exactly the right point of time. The ultimate goal is that every process should be provided with one part at a time, exactly when that part is needed.

The most important feature of the organizational set-up of the lean production system is the extensive use of multifunctional teams, which are groups of workers who are able to perform many different works. The teams are organized along a cell-based production flow system. The number of job-classifications also declines. Workers have received training to perform a number of different tasks, such as the statistical process control, quality instruments, computers, set-up performances, maintenance etc. In the lean production system responsibilities are decentralized. There is no supervisory level in the hierarchy. The multifunctional team is expected to perform supervisory tasks. This is done through the rotations of team leadership among workers. As a result, the number of hierarchical levels in the organization can be reduced. The number of functional areas that are the responsibility of the teams increases (Kumazawa and Yamada, 1989).

In a multifunctional set-up, it is vital to provide information in time and continuously in the production flow. The production system is changing and gradually adopting a more flexible system in order to adopt itself to changes in demand; which may reduce costs of production and increase efficiency. There are extensive usages of *Kaizen* activities and TQM and *Total Productive Maintenance* (TPM) to increase the effectiveness of corporate performances.

4. A Nation-Organization-Leader-Performance Model

Organizational culture is a product of a number of factors: NC, human resources practice, and leadership styles have prominent influences on it. These in turn, along with the OC, affect CPs. The proposed model, as in Fig. 1, is the synthesis of the ideas relating to this central theme. Figures 2 and 3 describe the proposed theoretical model relating OC and CPs. Corporate performances are measured in terms of the following criteria:

(a) customers' satisfaction;
(b) employee's satisfaction and commitment and
(c) contributions of the organization to the society and environment.

Most companies in Japan use these criteria to signify CPs (Basu, 1999; Nakane, 1970).

There are several latent or hypothetical variables, which jointly can influence the outcome. National culture is affected by two constructs:

MIVs and MAVs. Micro-values (MIVs) are composed of several factors (Kahle *et al.*, 1988). These are: (1) the sense of belonging; (2) respect and recognition from others and (3) a sense of life accomplishments and (4) self-respect.

Macro-values, according to the opinions of the executives of the Japanese, firms are religious values, habitual values, and moral values. There may be other MAVs, which are important for other nations like geography, racial origin, etc., but for the Japanese, these are not so important (Basu, 1999). Although moral and habitual values can be the results of religious values, it is better to separate out these three values as distinct. The moral and habitual values are different in Japan from other East Asian nations. Honor and "respect from others" are central to the Japanese psychology. Ritual suicides (*Harakiri*) are honorable acts for the Japanese, if they fail in some way to do their duty. Habitual values are extreme politeness on the surface, beautifications of everything, community spirit, and cleanness are unique in Japan, which are not followed in any other countries in Asia.

This is particularly true if we examine certain MEVs, which are neither MIV nor MAV values, but are derived from the NC. It is possible to identify five different MEVs, which are important outcomes of the Japanese NC. These are discussed as follows:

(a) Exclusivity or insider-outsider (*Uchi-Soto* in Japanese) psychology by which Japanese exclude anyone who is not an ethnic Japanese from social discourse; (It is different from color or religious exclusivities. For example, the Chinese or Koreans, who are living in Japan for centuries, are excluded from the Japanese social circles.)
(b) Conformity or the *doctrine of "nail that sticks up should be beaten down"* (*Deru Kuiwa Utareru* in Japanese) — deviations from the mainstream norms are not tolerated;
(c) Seniority system (*Senpai-Kohai* in Japanese) by which every junior must obey and show respects to the seniors;
(d) Collectivism in decision-making process ("*Hou-Ren-Sou*" system in Japanese) and
(e) Continuous improvements (*Keizen* in Japanese), which is the fundamental philosophy of the Japanese society.

These MEVs are exclusively Japanese, a reflection of the unique Japanese culture (Basu, 1999; Nakane, 1970) and is fundamental to the Japanese organizations.

National culture construct along with the MEVs affect both the organizational culture construct (OR), HRM system and leadership styles (LS) constructs. Japanese OR construct is affected by the NC, MEVs, and the HRM system. The similarities in OCs in different companies are due to the similarities in NC and MEV; the differences are due mainly to the differences in HRM. In Japan human resource practices vary from one company to another and as a result, OCs and performances vary. In Toyota, it is very consistent and strong culture with great emphasis on discipline, *Keizen*, TQM and *just in time* (JIT) production-inventory system. In many other companies in Japan, situations are not the same; as a result, OC differs.

Organizational culture is affected by the NC, MEVs, and HRM. If the company's leader is also the founder, then only the leadership style (LS) can affect the HRM and OC. Otherwise, when the leaders are professional managers trained and grown up within the company (In Japan and in Japanese companies abroad, it is out of practice to accept an outsider for the executive positions. All executives enter the company after their graduation and stay until they retire), appropriate LCs are developed by the OC and HRM.

Due to the collective decision-making process (*Hou-Ren-Sou*) leadership, a MEV in the model, cannot affect an organizational culture or the HRM system. It is very different from the Western (America, European, or Australian) companies where the leaders are hired from outside and are expected to be innovative regarding the OC and HRM system. This is quite alien to the Japanese culture. Leadership style in Japan is an outcome, not an independent variable as in the Western companies. Leadership style is affected by the NC, MEVs, OC, and the HRM.

Organizational culture (OC) is affected by four factors, according to this proposed model. These factors are: (a) stability; (b) flexibility; (c) internal focus and (d) external focus (Cameron and Quinn, 1999). These factors are universal, not restricted to the Japanese companies. However, how an OC is being affected by NC, MEVs, and HRM would vary from country to country.

Leadership style is a function of four factors. These factors are leader: (a) as a facilitator; (b) as an innovator; (c) as a technical expert and (d) as an effective competitor for rival firms (Cameron and Quinn, 1999). These four characteristics of a leader are universal, but the emphasis varies from country to country.

Finally, the corporate performance (CP) is affected by a number of influencing factors: (a) customers' satisfaction, (b) employees' satisfaction

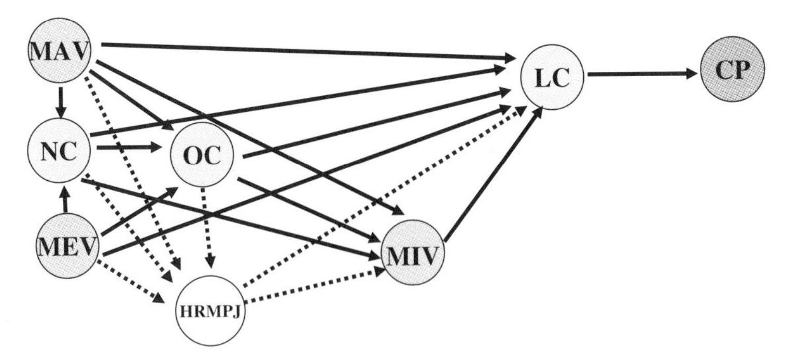

Where "NC" is "National Culture"; "OC" is "Organizational Culture"; "LC" is "Leader Culture"; "MAV" is "Macro-Values of Culture"; "MEV" is "Meso-Values of Culture"; "MIV" is "Micro-Values of Culture"; "HRMPJ" is "Human Resources Management Practices in Japan", and "CP" is "Corporate Performance".

Figure 2. Value–culture–performance: a thematic presentation.

and commitments, and (c) the contribution of the organization to the society. If we consider the contribution of the organization to society that is already taken into account by the customers' satisfaction and employees' satisfaction, then CP has these two important contributory factors.

Theoretical relationship between the observed variables and the underlying constructs are specified in an exploratory model, represented in Fig. 2. In Fig. 3, a more elaborate model, including the unique HRM system of the Japanese companies is described. Figure 3 represents a Linear Structural Relations (LISREL) version (Joreskog and Sorbom, 1999) of the model, which can be estimated using the methods of structural equation modeling (Raykov and Marcoulides, 2000).

5. Discussion

The proposed theory describes the relationship between NC and CPs through OC with the HRM system affecting the nature of the culture and leadership. These inter-relationships are important aspects of the Japanese corporate behaviors in a multinational setting and this proposed theory is an attempt to understand the Japanese corporate culture.

Analysis of the Japanese culture and organizations by outsiders so far suffers from a number of defects (Ikeda, 1987; Watanabe, 1998). They neither try to understand the effects of the Japanese value system on their organizations nor do they give any importance to the relationship between

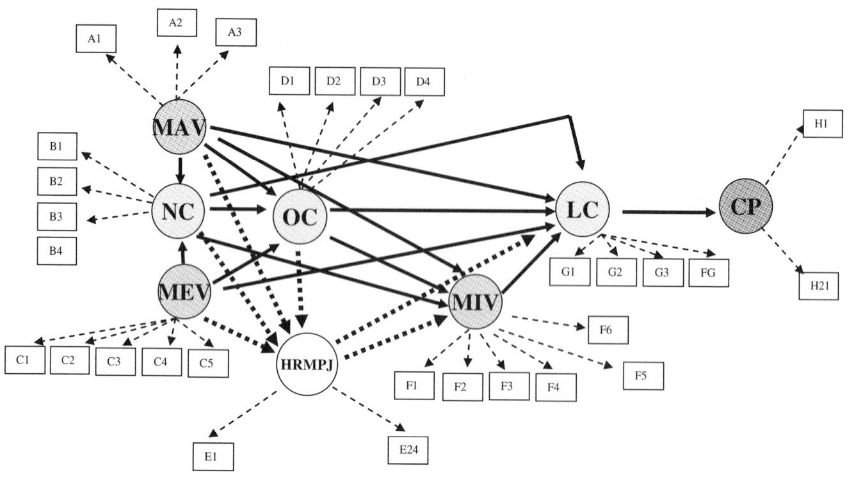

Where "NC" is "National Culture"; "OC" is "Organizational Culture"; "LC" is "Leader Culture"; "MAV" is "Macro-Values of Culture"; "MEV" is "Meso-Values of Culture"; "MIV" is "Micro-Values of Culture", and "CP" is "Corporate Performance"; A1 "Religious Values", A2 "Moral Values" A3 "Habitual Values" are indicators for MAV variable; B1 "Power Distance", B2 "Individualism/Collectivism", B3 "Masculinity/Femininity" are indicators for NC variable; C1 "Uchi/Soto Value", C2 "Deru Kuiwa Utareru Value", C3 "Senpai/Kohai Value", C4 "Hou-Ren-Sou Value", C5 "Kaizen Value" are indicators for MEV variable; D1 "Stability", D2 "Flexibility", D3 "External Focus", D4 "Internal Focus are indicators for OC variable"; E1 "On-Job-Training"; E2 "Emphasis on Personality of Recruits" are indicators for HRMPJ variable; F1 "Sense of Belonging to a Group Value", F2 "Respect and Recognition from Others Value", F3 "Sense of Life Accomplishment Value", F4 "Self-respect Value", F5 "Kangaekata Value", F6 "Ishiki Value" are indicators for MIV variable; G1 "Leader as Facilitator", G2 "Leader as Innovator", G3 "Leader as Technical Expert", G4 "Leader as Competitor" are indicators for LC variable; H1 "Customers Satisfaction", H2 "Employees satisfaction and Commitment" are indicators for CP variable.

Figure 3. Research model: Values-culture-performance (VCP): predicted paths between variables and indicators.

performance and culture. So far, the success of the Japanese companies was analyzed only in terms of their unique production and operations management system. However, the inter-relationship between the production and operations management system, which is a part of the OC, and the specific HRM system it requires for its proper implementation, was not given sufficient attention.

The typical example is the analysis of the failure of the implementations of Japanese operations management system in Britain by Oliver and Williamson (1992) or in China by Taylor (1999). They could not explain the reason for the failure, as they have not analyzed the Japanese OC and particularly the HRM system.

In a Japanese company, the LSs and the OC are designed by the HRM system; these are not coming from outside. An effective corporate performance is the result of these underlying determinants of OC and LSs. Thus, in a Japanese organization, LS is rooted in the HRM system, which emerges from the MEVs of the Japanese national culture.

Whether efficient CPs can be repeated in foreign locations depends on the transmission mechanism of the Japanese OC, which in turn, depends on the adaptations of the Japanese style HRM system. This can face obstacles due to the different MEVs in a foreign society and as a result, OC in a foreign location may be different from the OC of the company in Japan. Adaptation of the operations management system alone will not create an effective performance of a Japanese company in a foreign location. Creation of an appropriate HRM system and appropriate OC are essential for an effective performance.

6. Conclusion

A proposed theory of the relationship, among NC, OC, human resources practices, and CPs for the Japanese multinational companies, is narrated. Cultural differences are important and these factors affect every component of the company behaviors and performances. The question was raised regarding the factors, which are relatively more important than the others and whether it will be possible to manipulate these factors in order to increase the performances of the multinational companies. The proposed theory may provide an answer to these issues.

References

Aoki, M (1990). Towards an economic model of the Japanese firm. *Journal of Economic Literature* **28**, 1–27.

Axel, M (1995). Culture-bound aspects of Japanese management. *Management International Review* **35**(2), 57–73.

Barlett, CA and H Yoshihara (1988). New challenges for Japanese multinationals: is organization adaptation their Achilles Heel. *Human Resources Management* **27**(1), 19–43.

Basu, S (1999). *Corporate Purpose*. New York: Garland Publishers.

Calori, R and P Sarnin (1991). Corporate culture and economic performance: a French study. *Organization Studies* **12**(1), 49–74.

Cameron, KS and R Quinn (1999). *Diagnosing and Changing Organizational Culture: Based on the Competing Values Framework*. New York: Addison-Wesley.

Denison, DR and AK Mishra (1995). Toward a theory of organizational culture and effectiveness. *Organizational Science* **6**(2), March–April, 204–223.

Fujino, T (1998). How Japanese companies have globalized. *Management Japan* **31**(2), 1–28.

Hofstede, G (1980). *Culture's Consequences: International Differences in Work-related Values.* Beverly Hills, CA: Sage.

Hofstede, G (1985). The interaction between national and organisational value system. *Journal of Management Studies* **22**(3), 347–357.

Hofstede, G (1990). Measuring organizational cultures: a qualitative and quantitative study across twenty cases. *Administrative Science Quarterly* **35**, 286–316.

Hofstede, G (1993). Cultural constraints in management theories. *Academy of Management Executive* **7**(1), 81–94.

House, RJ *et al.* (1999). Cultural influences on leadership: Project GLOBE. In *Advances in Global Leadership*, W Moblet, J Gessner and V Arnold (eds.), Vol. 1. Greenwick: CT, JAI Press, 171–234.

Ikeda, K (1987). Nihon oyobi nihonjin ni tsuiteno imeji (Image of Japan and Japanese). In *Sekai wa Nihon wo Dou Miteiruka* (*How Does the World See Japan*). A Tsujimura, K Furuhata and H Akuto (eds.) Tokyo: Nihon Hyoronsa, pp. 12–31.

Imai, M (1986). *Kaizen: The Key to Japan's Competitive Success.* New York, NY: McGraw-Hill.

Joreskog, KG and D Sorbom (1999). *LISREL 8.30: User's Reference Guide.* Chicago: Scientific Software.

Kahle, LR, RJ Best and RJ Kennedy (1988). An alternative method for measuring value-based segmentation and advertisement positioning. *Current Issues in Research in Advertising* **11**, 139–155.

Kilman, R, MJ Saxton, and R Serpa (1985). Introduction: five key issues in understanding and changing culture, In *Gaining Control of the Organizational Culture*, R Kilman, M Gaxton and R Serpa (eds.), San Francisco: Jossey-Bass, pp. 1–16.

Kobayashi, N (1980). *Nihon no Takokuseki Kigyo* (Japanese multinational corporations). Tokyo: Chuo Keizaisha.

Kotter, JP and JL Heskett (1992). *Corporate Culture and Performance.* NY: The Free Press.

Kujawa, D (1979). The labour relations of US multinationals abroad: comparative and prospective views. *Labour and Society*, **4**, 3–25.

Kumagai, F (1996). *Nihon Teki Seisan Sisutemu in USA* (*Japanese Production System in USA*). Tokyo: JETRO.

Kumazawa, M and J Yamada (1989). Jobs and skills under the lifelong Nenko employment practice. In *The Transformation of Work.* S Wood (ed.). London: Unwin Hyman, 56–79.

Marcoulides, GA and RH Heck (1993). Organization culture and performance: proposing and testing a model. *Organization Science* **14**(2), May, 209–225.

Morishima, M (1996). Renegotiating psychological controls, Japanese style. In *Trends in Organizational Behavior.* CL Cooper (ed.), Vol. 3. New York: John Wiley & Sons Ltd., 149–165.

Morita, A (1992). A moment for Japanese management. *Japan-Echo* **19**(2), 1–12.

Miyajima, H (1996). *The evolution and change of contingent governance structure in the J-firm system: An approach to presidential turnover and firm performance.* Working Paper No. 9606. Institute for Research in Contemporary Political and Economic Affairs. Tokyo: Waseda University.

Nakane, G (1970). *Japanese Society.* Harmondsworth: Penguin.

Oliver, N and B Wilkinson (1992). *The Japanization of British Industry*. Cambridge, Massachussets: Blackwell.

Ouchi, W (1981). *Theory Z: How American Business can Meet the Japanese Challenge*. Reading, MA: Addison-Wesley.

Raykov, T and GA Marcoulides, (2000). *A First Course in Structural Equation Modeling*. London: Lawrence Erlbaum.

Schein, EH (1997). *Organizational Culture and Leadership*. San-Francisco: Jossey-Bass Publishers.

Shimada, H (1993). Japanese management of auto-production in the United States: An overview of human technology in international labour organization. In *Lean Production and Beyond*, Geneva: ILO, 23–47.

Taylor, B (1999). Patterns of control within Japanese manufacturing plants in China: doubts about Japanization in Asia. *Journal of Management Studies* **36**(6), 853–873.

Watanabe, S (1998). Manager-subordinate relationship in cross-cultural organizations: the case of Japanese subsidiaries in the United States. *International Journal of Japanese Sociology* **7**, 23–43.

Topic 3

Health Service Management Using Goal Programming

Anna-Maria Mouza
Institute of Technology and Education, Greece

1. Introduction

Private production units usually operate on the basis of profit maximization. However, to face competition from similar units and to be in line with some major socio-economic factors, some decision makers are forced to relax this basic objective. In order to present propositions for optimal management decisions, one should combine the priorities of the decision maker with the technical and socio-economic factors, which are involved in the operating process in the best possible way. The production unit considered in this paper is a relatively small (60 beds) private clinic in northern Greece, which is similar to the orthopedic department of a hospital studied and described elsewhere (Mouza, 1996).

However, in this particular case, there is no out-patient services. To facilitate the presentation, I consider a five-year operational plan where the personnel claims, the manpower working pattern, the various expenses, together with the profit maximization target, and the expected number of patients are described in detail. The necessary projections are based on reliable techniques presented elsewhere (Mouza, 2002; 2006).

After setting the priorities of the various targets, I formulate a proper "goal programming" problem using deviational variables (d_i^- and d_i^+) and obtained the first-run solution. Then, I proceed to the goal attainment evaluation, which dictated a re-adjustment of the priorities. The second-run solution indicates that a slower acceleration of the profit maximization rate is preferable, to avoid over-charging the patients, at the same time satisfying most of the requirements for personnel management. Furthermore, if the decision maker's preference is relatively rigid, a lower rate of the increase of salaries of the employees should be adopted. All computational results are presented at the end of the relevant section.

2. The Details of the Private Clinic in Relation to the Five-Year Plan

In Table 1, I present the composition of the personnel, and their salary at the base year (t_0). Note that the 5-year planning horizon, refers to the time-periods (years) t_1, t_2, t_3, t_4, and t_5.

In columns 4 and 5 of Table 1, the total amount of the wages for all employees of the same group, for the initial and final year is stated. Note that the wages are an average of the salaries earned by each person in the individual group. For each group, an average percent of annual salary

Table 1. Personnel of the orthopedic clinic and their wages.

Groups	Position of the group members	Members at each group	Annual salary at year t_0 for all group members (in thousand €)	Salary at final year t_5, for all group members (in thousand €)
1	Orthopedic surgeons, anaesthiologists and microbiologists	10	260	356.222
2	Pharmacist	1	24.7	33.054
3	Operating-room nurses	3	54.6	68.04
4	Nurses plus laboratory staff	14	163.8	199.29
5	Axial-tomography and X-ray technicians	3	58.5	74.307
6	Clinic manager	1	27.3	36.190
7	Secretary and administrative service staff	3	32.76	39.476
8	Office clerks	4	41.6	49.647
9	Cooking and food service staff	4	42.64	50.643
10	Other staff (guard, bearers maintenance, cleaning service etc.)	10	101.4	119.851

increase, say R_i, is considered. It should be noted that in some groups, this rate is well above the rate of inflation growth, and the rate stated in the collective wage and salary agreements for each employees group. These rates of annual salary increase for all groups are presented in Table 3. Denoting by $A_0(i)$ the annual salary of all members of group i at the base year t_0, and the final annual salary at year t_5 for all members of the same group, denoted by $A_n(i)$, is computed from:

$$A_n(i) = A_0(i)\left(1 + \frac{R_i}{100}\right)^5. \tag{1}$$

Thus for group 1, where $R_1 = 6.5\%$, the final salary for all group members will be:

$$A_n(1) = 260(1 + 0.065)^5 = 356.222.$$

It should be noted, that these rates of salary increases have been formed in accordance to the personnel claims. Other additional expenses for the final year of of the planning period are presented in Table 2.

On the top of Table 2, the forecast regarding the expected number of patients for the final year is stated. It should be pointed out that not all patients necessarily need an axial-tomography and/or X-ray treatment. That is why the average per patient cost stated in Table 2 seems to be relatively low.

The interactive voice response (IVR) systems which are effectively used in healthcare and hospitals are described by Mouza (2003). It should be noted that replacement expenses together with expenses for obtaining new devices, seems more reasonable to be equally distributed over the five years of the planning period. Needless to say, that in case of replacements, the estimated salvage on old equipment has to be subtracted from the relevant cost. Thus, at the final year, only one fifth of such expenses is stated. For the case under consideration, the replacement refers to the server of the local area computer network (LAN), together with two stations. Their net cost sum up to €2000, which has been equally spread over the five years of the planning period.

From Table 1, I computed the average salary per employee at the final year which, when multiplied by the group size gives the total salary for the corresponding group, as it can be verified from Table 3.

It can be seen from Table 3, that $X_i (i = 1, \ldots, 10)$ denotes the size (i.e., the members) of group i and c_i, the average salary of each group member. Thus, the total salary of all the members of group 1, for instance, can also

Table 2. Forecasted number of patients and various expenses for the final year.

Patient expenses		Forecasted average per patient, at the final period t_f (€)
Forecasted number of patients for the period (year) $t_5 = 3625$		
Axial-tomography		9.1
X-ray		5.7
Medical supplies		59.4
Administrative and miscellaneous		32.1
Other expected expenses	Total (€)	Amount for the final year t_5
Installation of IVR system	3000	600
Server and computers replacement	2350 (salvage value: 350)	400
Personnel insurance (19% of total salaries)		Endogenous (decision variable X_{39})
Laboratory expenses		60,000
Fixed expenses plus depreciation		180,000
All other expenses		80,000

be computed from:

$$c_1 \times X_1 = 35622.2 \times 10 = 356222.$$

If I denote by X_{10+i}, the salary expenses of group i, then I may write

$$X_{10+i} = c_i \times X_i \quad (i = 1, \ldots, 10). \tag{2}$$

so that the total salary expenses for the final year can be determined by evaluating the summation:

$$\sum_{i=11}^{20} X_i. \tag{3}$$

Table 3. Employees' salary at the initial and final year of the planning period.

Groups (i)	Number of group members (X_i)	Annual salary at year t_0, for all the members of group i (in thousand €)	Annual rate of increase, R_i (in %)	Salary at final year t_f, for all the members of group i (in €)	Average per person at each group (c_i)
1	10	260	6.5	356,222	35,622.2
2	1	24.7	6	33,054	33,054
3	3	54.6	4.5	68,040	22,680
4	14	163.8	4	199,290	14,235
5	3	58.5	4.9	74,307	24,769
6	1	27.3	5.8	36,190	36,190
7	3	32.76	3.8	39,476	13,158.666
8	4	41.6	3.6	49,647	12,411.75
9	4	42.64	3.5	50,643	12,660.75
10	10	101.4	3.4	119,851	11,985.10

Another point of interest refers to the average working hours of the members of each group. The relative figures are analytically presented in Table 4.

It can be easily verified that in Table 4, column 5 is obtained by multiplying the elements of column 4 by 52.

Denoting by $X_{22+i} (i = 1, \ldots, 10)$ the total average working hours per year for all the group members, then I may write

$$X_{22+i} = w_i \times X_i \quad (i = 1, \ldots, 10). \tag{4}$$

3. The Nature of the Problem

Given all the information presented in Tables 1–4, the resultant problem is, how to formulate a feasible operating plan, which should combine the desired salary increases and the full utilization of working hours, the expenses incurred, together with the desired gross profit at the final year set by the decision maker, in the best possible way with the minimum sacrifice. This problem can be answered by the goal programming method, originally developed by Charnes and Cooper (1961). It should

Table 4. Working hours of each group personnel.

Groups (i)	Number of group members	Hours per week (average), for each member of group i (in thousand €)	Total hours per week for all the members of group i	Total hours per year, for all the members of each group	Average per person at each group (w_i)
1	10	65	650	33,800	3380
2	1	45	45	2340	2340
3	3	65	195	10,140	3380
4	14	40	560	29,120	2080
5	3	40	120	6240	2080
6	1	45	45	2340	2340
7	3	40	120	6240	2080
8	4	40	160	8320	2080
9	4	30	120	6240	1560
10	10	30	300	15,600	1560

be recalled that goal programming is a mathematical programming technique with the capability of handling multiple goals given certain priority structure. Furthermore, a goal programming model may be composed of non-homogeneous units of measure, which is the case under consideration. In brief, the main task here refers to the compensation of total expenses with the decision maker's requirement regarding gross profits, without any change in the operating pattern of the clinic, as far as the manpower level is concerned.

3.1. *Specification of the Problem Decision Variables*

- The first 10 decision or choice variables $X_i (i = 1, \ldots, 10)$ are already defined in Table 3 and refer to the number of personnel in group i.
  - — Variables $X_{11} \ldots X_{20}$, stand for the salary cost of each group.
  - — Variable X_{21} denotes total salary cost at the final year.
  - — Variables $X_{22} \ldots X_{31}$ refer to the total working hours, observed in the 5th column of Table 4.

- Variables $X_{32} \ldots X_{35}$ refer to the average cost per patient, as it is stated in the first four rows of Table 2.
  - — Variable X_{36} denotes total per patient expenses.
  - — Variables X_{37} and X_{38} refer to the equipment purchase and replacement.
- Variable X_{39} stands for the personnel insurance expenses.
  - — Variable X_{40} refers to the depreciation, laboratory, fixed and other expenses.
- Variable X_{41} is used to sum up all the above expenses (i.e., $X_{37} + X_{38} + X_{39} + X_{40}$)
- X_{42}, which also is a definition variable, is used to denote total operating cost for the final year t_f (i.e., t_5) of the planning period.
- Finally, variable X_{43} denotes the average charge per patient, so that the desired gross profit would not be less than €750,000.

3.2. *Formulation of the Constraints (Goals)*

To facilitate the presentation, I classified the goals into the following seven categories.

3.2.1. *Personnel Employed*

The operation pattern of the clinic, regarding the personnel employed is presented in Table 1. Given that, the decision maker does not want to alter this pattern, I must first introduce the following 10 constraints (goals)

$$X_i + d_i^- - d_i^+ = b_i \quad (i = 1, \ldots, 10) \tag{5}$$

where $b_1 = 10$, $b_2 = 1$, $b_3 = 3$, $b_4 = 14$, $b_5 = 3$, $b_6 = 1$, $b_7 = 3$, $b_8 = 4$, $b_9 = 4$ and $b_{10} = 10$, are the desired targets, regarding the members of each group as shown in Tables 1, 3 and 4.

It is useful to recall that d_i^- and d_i^+ denote the so-called deviational variables from the goals, which are complementary to each other, so that $d_i^- \times d_i^+ = 0$. If the ith goal is not completely achieved, then the observed slack will be expressed by d_i^-, which represents the negative deviation from the goal. On the other hand, if the ith goal is over achieved, then d_i^+ will take a non-zero value. Finally, if this particular goal is exactly achieved, then both d_i^- and d_i^+ will be zero.

3.2.2. *Personnel Claims for Salary Increase*

According to Eq. (2), I have to introduce the following 10 constraints in order to define the total salary expenses for each group.

$$X_{10+i} - c_i X_i + d^-_{10+i} - d^+_{10+i} = 0, \quad (i = 1, \ldots, 10). \tag{6}$$

It is recalled that coefficients c_i are presented in Table 3. An additional variable X_{21} is introduced to sum up salary expenses for all personnel at the final year. The computation of this total is based on Eq. (3).

$$X_{21} - \sum_{i=11}^{20} X_i + d^-_{21} - d^+_{21} = 0. \tag{7}$$

3.2.3. *Complete Utilization of Personnel Capacity in Terms of Working Hours*

The forecasted number of patients for the final year is not exceeding the clinic capabilities, regarding the existing infrastructure. Furthermore, the decision maker does not want to alter the existing operation pattern, since he considers that the present personnel manpower potential will be adequate to provide satisfactory service to the number of patients forecasted, provided that the under utilization of the regular working hours will be avoided. Hence, I introduced the following constraints in accordance to Eq. (4), which on the other hand may be regarded as a measure to provide job security to all personnel by avoiding underutilization of their regular working hours as shown in Table 4.

$$X_{21+i} - w_i X_i + d^-_{21+i} - d^+_{21+i} = 0, \quad (i = 1, \ldots, 10). \tag{8}$$

Note, the co-efficients w_i, are also presented in Table 4.

3.2.4. *Patient expenses*

The axial-tomography expenses per patient can be expressed as:

$$X_{32} + d^-_{32} - d^+_{32} = 9.1. \tag{9a}$$

X-ray expenses per patient:

$$X_{33} + d^-_{33} - d^+_{33} = 5.7. \tag{9b}$$

Medical expenses per patient:

$$X_{34} + d^-_{34} - d^+_{34} = 59.4. \tag{9c}$$

Administrative and miscellaneous expenses per patient:

$$X_{35} + d_{35}^{-} - d_{35}^{+} = 32.1. \tag{9d}$$

Variable X_{36} depicts total per patient expenses at the final year.

$$X_{36} - (X_{32} + X_{33} + X_{34} + X_{35}) + d_{36}^{-} - d_{36}^{+} = 0. \tag{9e}$$

3.2.5. *Other Expected Expenses*

Installation of the IVR system.

$$X_{37} + d_{37}^{-} - d_{37}^{+} = 600. \tag{10a}$$

Computers and server replacement.

$$X_{38} + d_{38}^{-} - d_{38}^{+} = 400. \tag{10b}$$

The insurance funds.

From Table 2, we see that the average expenses for personnel insurance amounts to 19% of the total employees' salary. This is represented by the following constraint, taking into account the variable X_{21}.

$$X_{39} - 0.19X_{21} + d_{39}^{-} - d_{39}^{+} = 0 \tag{10c}$$

Depreciation, laboratory, fixed and other expenses

$$X_{40} + d_{40}^{-} - d_{40}^{+} = (60{,}000 + 180{,}000 + 80{,}000). \tag{10d}$$

To sum all other expenses, I introduce the following relation.

$$X_{41} - (X_{37} + X_{38} + X_{39} + X_{40}) + d_{41}^{-} - d_{41}^{+} = 0. \tag{10e}$$

3.2.6. *Total Expenses*

It is already mentioned that I introduce the variable X_{43} in order to sum up total expenses, in the following way.

$$X_{42} - (X_{21} + 3625X_{36} + X_{41}) + d_{42}^{-} - d_{42}^{+} = 0. \tag{11}$$

3.2.7. *Total Gross Profit*

The gross profit achievement as set by the decision maker, may be introduced in the following way.

$$3625X_{43} - X_{42} + d_{43}^{-} - d_{43}^{+} = 750{,}000 \tag{12}$$

where X_{43}, as it was mentioned earlier, represents the adequate charge per patient, in order to satisfy Eq. (12). It is recalled that the coefficient 3625, which appears in Eq. (11) too, is the stated forecast, regarding the expected number of patients for the final year t_f.

With the above formulation, we see that the number of constraints is equal to the number of choice variables (43).

3.3. *Priority of the goals — The objective function*

It is obvious that the decision maker, apart from determining the goals of the five-year plan, also has to specify the priorities (Gass, 1986), denoted by P_i, in order to accomplish the optimum allocation of resources for achieving these goals in the best possible manner. One further point, that has to be considered in the formulation of the model, is the possibility of weighing the deviational variables at the same priority level (Gass, 1987). The list of decision maker's goals is presented below in descending order of importance.

1. To retain the present operating pattern of the clinic.

This implies that no alterations are desired regarding the number and composition of the existing personnel (P_1). Different weights have been assigned to the various personnel groups.

2. To achieve the desired level of gross profits. Also, to provide reserves for the personnel insurance, to cover all expenses for equipment replacements and installation of a new IVR system (P_2).
3. To avoid underutilization of the personnel regular working hours, providing at the same time job security to all employees (P_3).
4. To provide the funds necessary to cover all patient's expenses (P_4).
5. To provide adequate funds for covering laboratory and fixed expenses, depreciation etc. (P_5).
6. To provide the wage increase, claimed by the personnel at the various groups (P_6).

In this case, weights are used to discriminate the salary increase of each group. Heavier weights have been assigned to groups with lower salaries.

According to the above ranking of the goals priority and taking into account the relations for the corresponding totals, the following objective function is formulated.

$$\begin{aligned}\min z = {} & (10P_1d_1^- + 9P_1d_2^- + 8P_1d_3^- + 7P_1d_4^- + 6P_1d_5^- + 5P_1d_6^- \\ & + 4P_1d_7^- + 3P_1d_8^- + 2P_1d_9^- + P_1d_{10}^-) + (P_2d_{37}^- + P_2d_{38}^- \\ & + P_2d_{39}^- + 3P_2d_{42}^- + P_2d_{43}^-) + P_3\sum_{i=22}^{31} d_i^- + P_4\sum_{i=32}^{36} d_i^- \\ & + (P_5d_{40}^- + P_5d_{41}^-) + (P_6d_{21}^- + 2P_6d_{11}^- + 2P_6d_{12}^- + 3P_6d_{13}^- \\ & + 4P_6d_{14}^- + 5P_6d_{15}^- + 6P_6d_{16}^- + 7P_6d_{17}^- + 8P_6d_{18}^- \\ & + 9P_6d_{19}^- + 10P_6d_{20}^-). \end{aligned} \tag{13}$$

4. Initial Results

With this specification of the goal programming model, I obtain the results as given on the left side of Table 5 (solution I). It is easily verified that all goals are achieved. Hence, this pattern dictated by the optimal solution is readily applicable. It should be pointed out that according to this solution, an average rate of total salary increase of about 4.9% is realized, whereas the mean salary per worker increases from 15,232.1 to 19,372.1. In fact, this average rate r (i.e., 4.9) has been computed from Eq. (1) in the following way, using natural logs.

$$\frac{\ln(x)}{5} = \ln(1+r), \quad \text{where } x = \frac{1{,}026{,}720}{807{,}300}$$

and $r = (y - 1) \times 100$, where $y = e^Z$ and $Z = \frac{\ln(x)}{5}$

However, the charge per patient (~740€) is considered being higher, when compared to the average of the local market, affecting thus the competition level of the clinic.

5. Imposition of a New Constraint

Given the social character of health service, which has to be combined with the fact that the unit under consideration must be competitive, it is necessary to impose the following constraint.

$$X_{43} + d_{44}^- - d_{44}^+ = 700. \tag{14}$$

Thus, the total number of constrains has become 44. It should be noted that minimizing d_{44}^{+} in the objective function, an upper limit of €700 is set to the average charge per patient. Hence, I added in the objective as in Eq. (13) the following term

$$2P_2d_{44}^{+}$$

which implies that the restriction regarding the maximum average charge per patient has been considered as a second priority goal, properly weighed.

5.1. *Second-Run Results*

With the new constraint, the solution obtained is presented on the right side of Table 5 (solution II). I observe that constraint Eq. (14) is binding, and in order to achieve the level of gross profit set by the decision maker, the solution dictates that the total salary expenses must be reduced by €118,180.125, which is the value of the deviational variable d_{21}^{-}, shown in Eq. (7). Evaluating the results of the second run, one may argue as to

Table 5. Solution I (left) and solution II (right).

No.	Variable name	Value	No.	Variable name	Value
The simplex solution (Number of patients = 3625)					
87	×1	10.000	89	×1	10.000
88	×2	1.000	90	×2	1.000
89	×3	3.000	91	×3	3.000
90	×4	14.000	92	×4	14.000
91	×5	3.000	93	×5	3.000
92	×6	1.000	94	×6	1.000
93	×7	3.000	95	×7	3.000
94	×8	4.000	96	×8	4.000
95	×9	4.000	97	×9	4.000
96	×10	10.000	98	×10	10.000
97	×11	356,222.000	99	×11	356,222.000
98	×12	33,054.000	100	×12	33,054.000
99	×13	68,040.000	101	×13	68,040.000
100	×14	199,290.000	102	×14	199,290.000
101	×15	74,307.000	103	×15	74,307.000
102	×16	36,190.000	104	×16	36,190.000

(*Continued*)

Table 5. **(*Continued*)**

No.	Variable name	Value	No.	Variable name	Value
The simplex solution (Number of patients = 3625)					
103	×17	39,476.000	105	×17	39,476.000
104	×18	49,647.000	106	×18	49,647.000
105	×19	50,643.000	107	×19	50,643.000
106	×20	119,851.000	108	×20	119,851.000
107	×21	1,026,720.000	21	d_{21}^{-}	118,180.125
108	×22	33,300.000	110	×22	33,800.000
109	×23	2340.000	111	×23	2340.000
110	×24	10,140.000	112	×24	10,140.000
111	×25	29,120.000	113	×25	29,120.000
112	×26	6240.000	114	×26	6240.000
113	×27	2340.000	115	×27	2340.000
114	×28	6240.000	116	×28	6240.000
115	×29	8320.000	117	×29	8320.000
116	×30	6240.000	118	×30	6240.000
117	×31	15,600.000	119	×31	15,600.000
118	×32	9.100	120	×32	9.100
119	×33	5.700	121	×33	5.700
120	×34	59.400	122	×34	59.400
121	×35	32.100	123	×35	32.100
122	×36	106.300	124	×36	106.300
123	×37	600.000	125	×37	600.000
124	×38	400.000	126	×38	400.000
125	×39	195,076.797	127	×39	172,622.578
126	×40	320,000.000	128	×40	320,000.000
127	×41	516,076.812	129	×41	493,622.562
128	×42	1,928,134.375	130	×42	1,787,500.000
129	×43	738.796	131	×43	700.000
			109	×21	908,539.875

whether this is an optimal solution, from the socio-economic point of view, since any reduction has to directly affect the employees salary solely. On the other hand, the decision maker is not willing to entirely undertake this reduction (118,180.125), decreasing gross profits by almost 16%. With all these in mind, I present the following proposition, which may be considered of particular importance.

6. Some Alternatives

After proper negotiations, the decision maker and the employees have to agree on the proportion, say a (where $0 < a < 1$), of the deficit that will be subtracted from the total employee's salary. Let us assume that both sides agree for $a = 0.5$. In this case, the total salary of all employees will be reduced by $0.5 \times 118{,}180.125 = 59{,}090.0625$, and would affect their earnings proportionally. A crucial point of particular importance, is that special care should be taken for retaining a fair wage policy. This implies that lower salary groups should undergo lesser decrease, compared to high salary groups. In any case, the mean annual rate of salary increase for each group after the reduction mentioned above should not be less than 3.1% over the planning period, which is an estimate of the expected annual rate of inflation increase. In fact, the latter requirement affects the value of "a".

All these restrictions call for the construction of a composite index, in order to distribute the salary restraint among different groups in the desired and acceptable way. For this reason, the initial rates of salary increase for each group, presented in Table 3, have been ranked properly as it is shown in Table 6. Finally, after taking into account, the initial rates of salary

Table 6. The distribution of salary reduction among groups, using the composite index.

Groups (i)	Number of group members (X_i)	Annual rate of initial salary increase r_i (in %)	Ranking for salary decrease	A composite index to serve as the rate of initial salary decrease	Corresponding amount (in €)
1	10	6.5	10	0.627689	37,090.172
2	1	6	9	0.048387	2859.187
3	3	4.5	6	0.049801	2942.746
4	14	4	5	0.108050	6384.697
5	3	4.9	7	0.069093	4082.711
6	1	5.8	8	0.045522	2689.870
7	3	3.8	4	0.016266	961.173
8	4	3.6	3	0.014535	858.898
9	4	3.5	2	0.009610	567.861
10	10	3.4	1	0.011047	652.748
Total				1	59,090.06

Table 7. Salary at the initial and final years together with the annual rates of increase.

Groups (i)	Salary at the period t_0 (€)	Initially claimed salary for the final year (€)	Amount of reduction	Salary to be realized at the final year (€)	Actual annual rate of salary increase (%)	New coefficients c_i
1	260,000	356,222	37,090.172	319,131.81	4.1836	31,913.181
2	24,700	33,054	2859.187	30,194.81	4.0991	30,194.81
3	54,600	68,040	2942.746	65,097.25	3.5795	21,699.083
4	163,800	199,290	6384.697	192,905.3	3.3252	13,778.95
5	58,500	74,307	4082.711	70,224.29	3.7209	23,408.096
6	27,300	36,190	2689.870	33,500.13	4.1782	33,500.13
7	32,760	39,476	961.173	38,514.83	3.2897	12,838.278
8	41,600	49,647	858.898	48,788.10	3.2391	12,197.025
9	42,640	50,643	567.861	50,075.14	3.2669	12,518.785
10	101,400	119,851	652.748	119,198.25	3.2872	11,919.825

increase for each group, the number of employees per group together with the ranking scores, I computed the index presented in the 4th column of Table 6. Multiplying the elements of this column by 59,090.0625, we get the entries of the 5th column.

From the last column of Table 6, I computed the end period salary of each group and the corresponding annual rate of increase, which are presented in Table 7.

From Table 7, we can see that the requirement for the annual rate of salary increase at each group to exceed 3.1%, is fully satisfied. In an opposite situation however, one has to reduce the value of "*a*" by, say, 0.05 and to repeat all calculations. It should be also noted that with the allocation proposed so far, total initial salary, spending will increase by an annual rate of 3.69% (from 807,300 to 967,629.9) and the annual wage per capita will increase by 19.86% (from 15,232.1 to 18,257.2). It is clear that all employees will be better off with smaller value of *a*. The composite index and all other entries of Tables 6 and 7 can be computed with the following code segment written in Visual Fortran.

```
MODULE data_and_params
! Goups=number of groups   ! Planning_hor=Planning horizon (years)
! R = Initial rate of salary increase at each group (vector)
! Members    =  Number of employees at each group (vector)
           ! Salary_in =  Initial employees salary at each group (vector)
! Salary_f   =  Final claimed salary for each group (vector)
! Deficit    =  The value of deviational variable
! a          = Value of coefficient a   (0<a<1)
  IMPLICIT  NONE
   INTEGER, PARAMETER          :: groups= 10, planning_hor=5, out = 8
   REAL, DIMENSION(groups)     :: R, Salary_in, Salary_f
   INTEGER, DIMENSION(groups) :: Members
   REAL                        :: Total_Deficit, a
! To facilitate the presentation data are given directly here
   DATA R          / 6.5, 6.0, 4.5, 4.0, 4.9, 5.8, 3.8, 3.6, 3.5, 3.4/
   DATA Members    / 10; 1, 3, 14, 3, 1, 3, 4, 4, 10 /
   DATA Salary_in / 260000., 24700., 54600., 163800., 58500., 27300. &
                    & ,32760., 41600., 42640., 101400./
   DATA Salary_f   / 356222., 33054., 68040., 199290., 74307., 36190. &
                    & ,39476., 49647., 50643., 119851./
   DATA Total_Deficit, a /118180.125, 0.5/
 END MODULE data_and_params
!
 PROGRAM composite_index
  USE Dflib
  USE data_and_params
  IMPLICIT NONE
  INTEGER     i, j, Employees
  REAL DIMENSION(groups)   :: Y, X, rate, Rsmall, sortR, Scores, newrate
  REAL                        Logvector(groups),Salary(groups), Deficit, s
! Open output file
   OPEN(out, FILE='Composite_Index.out', STATUS='UNKNOWN')
```

```
   WRITE(out,*)' Composite index and the final rate of salary increase'
   WRITE(out,'(/)')
   Deficit=Total_Deficit*a  ;  sortR=R  ; Salary=Salary_f/members
! Perform ranking operation
   CALL SORTQQ(LOC(sortR),groups,SRT$REAL4)
   DO i=1,groups
      s=R(i)
      DO j=1,groups
         IF(sortR(j) == s) scores(i)= j
     ENDDO  ;  ENDDO
! Compute composite index
   Rsmall=R/100.        ;  Y=Rsmall*Salary   ;  s=SUM(Y)
   Employees=SUM(Members)
   rate=(y*100)/s/100.   ;  X=rate*deficit/Employees
   s=SUM(X)
   rate=(x*100.)/s/100.  ;  s=SUM(X)      ; X=rate*Members
   s=SUM(X)
   Rate=(X*100.)/s/100.  ;  s=SUM(Rate)   ;  X=rate*scores
   s=SUM(X)
   rate=(X*100)/s/100.   ;  s=SUM(rate)    ;  X=Deficit*rate
! Compute revised annual rates of salary increase
   Y=Salary_f-X
   Logvector=(LOG(Y/Salary_in))/ planning_hor
   newrate=(EXP(logvector) -1.)*100.
   WRITE(out,*) '  Index      Reduction    Final Salary   New rate'
   DO i=1,groups
      WRITE(out,'(1X,F9.6,2X,F10.3,4X,F10.2,6X,F6.4)') Rate(i)&
                  &, X(i), Y(i), newrate(i)
   ENDDO
 END
```

7. Final Results

In the last column of Table 7, the revised coefficients c_i are stated. These new coefficients are to be considered in Eq. (2), and in the corresponding restrictions seeing in Eq. (6) for the final run. Also, the objective in Eq. (13) has been slightly reformed as follows:

$$\begin{aligned}\min z = {} & (10P_1d_1^- + 9P_1d_2^- + 8P_1d_3^- + 7P_1d_4^- + 6P_1d_5^- + 5P_1d_6^- \\ & + 4P_1d_7^- + 3P_1d_8^- + 2P_1d_9^- + P_1d_{10}^-) + (P_2d_{37}^- + P_2d_{38}^- \\ & + P_2d_{39}^- + 3P_2d_{42}^- + 2P_2d_{44}^+) + P_3\sum_{i=22}^{31} d_i^- + P_4\sum_{i=32}^{36} d_i^- \\ & + (P_5d_{40}^- + P_5d_{41}^-) + (P_6d_{21}^- + 2P_6d_{11}^- + 2P_6d_{12}^- \\ & + 3P_6d_{13}^- + 4P_6d_{14}^- + 5P_6d_{15}^- + 6P_6d_{16}^- + 7P_6d_{17}^- + 8P_6d_{18}^- \\ & + 9P_6d_{19}^- + 10P_6d_{20}^-) + P_7d_{43}^-.\end{aligned}$$

The solution obtained with the re-stated restrictions satisfies all requirements. Now, total expenses sum up to €1,857,817.125. It has to be underlined, however, that since the value of the deviational variable d_{43}^- is 70,317.2, it is implied that total gross profits undergo a reduction of about 9.38% (from 750,000 to 679,682.8), which is a bit higher when compared to the corresponding percentage (~6%) of total salary reduction. Nevertheless, the results provide a sound basis to achieve an optimal solution, in the sense that it is acceptable by all people involved without any violations of the existing rules.

8. Conclusions

Many clinics like the one presented here benefit the patients, the health service in a convenient way, and help attain consistency and continuity of treatment. These achievements are heavily dependent on a successful budget planning. The characteristics of a model of this type for a clinic are based upon many factors, such as the type of the clinic, the location, the size, the medical specialty, etc. Hence, it is difficult to design a general model that can be applied to all types of clinics. However, once a budget planning model has been developed, it can be easily modified at a later stage to fit many other types of clinics. With this in mind, I consider in this paper an orthopedic clinic and implemented a model designed for a five-year planning period, using the goal programming approach of aggregate budget planning for this particular health care unit. To start with, the goals and the relevant priorities are set in advance, to formulate the corresponding goal programming model. The solution obtained is operational in the sense that, through the analysis of resource requirements, the acceptable charge per patient, the desired level of gross profits, the claimed rate of salary increases by the employees, and the trade-offs among the set of goals are achieved with a minimum sacrifice. The model shows that this sacrifice is negotiable between different sides, so that the business manager can finally establish an aggregative budget planning with the best perspectives, since the realization of a fair wage policy can be obtained through using the composite index, introduced in this paper.

References

Charnes, A and W Cooper (1961). *Management Models and Industrial Applications of Linear Programming* (Vols. 1 and 2). New York: Wiley.

Everett, JE (2002). A decision support simulation model for the management of an elective surgery waiting system. *Health Care Management Science* **5**, 89–95.

Flessa, S (2000). Where efficiency saves lives: a linear programme for the optimal allocation of health care resources in developing countries. *Health Care Management Science* **3**, 249–267.

Gass, SI (1986). A process for determining priorities and weights for large scale linear goal programming. *Journal of Operational Research Society* **37**, 779–785.

Gass, SI (1987). The setting of weights in linear goal programming. *Computers and Operations Research* **14**, 227–229.

Mouza and Anna-Maria (1996). An economic approach of the Greek health sector, based on domestic data. Modeling the hospital operation process. Ph.D. Theses, Greece: Aristotle University of Thessaloniki.

Mouza and Anna-Maria (2002). Estimation of the total number of hospital admissions and bed requirements for 2011: the case for Greece. *Health Service Management Research* **15**, 186–192.

Mouza and Anna-Maria (2003). IVR and administrative operations in healthcare and hospitals. *Journal of Healthcare Information Management* **17**(1), 68–71.

Mouza and Anna-Maria (2006). A note regarding the projections of some major health indicators. *Quality Management in Health Care* **15**(3), 184–199.

Ogulata, SN and R Erol (2003). A hierarchical multiple criteria mathematical programming approach for scheduling surgery operations in large hospitals. *Journal of Medical Systems* **27**(3), 259–270.

Chapter 5
Modeling National Economies

Topic 1

Inflation Control in Central and Eastern European Countries

Fabrizio Iacone and Renzo Orsi
University of Bologna, Italy

1. Introduction

The recent accession to the European Union (EU) of 10 new members officially opened the agenda of their participation in the European Monetary Union (EMU) too.

Although, the accession of the new European partners to the euro-area is formally analyzed, keeping the Maastricht criteria as the main reference, these do not necessarily reflect the effective integration and convergence of the economies. With this respect, we think it is important to study if and how the Eurosystem can successfully control inflation after the extension of the EMU: we then investigated the monetary policy transmission mechanism in the Central and Eastern European Countries (CEECs) to see how compatible it is with one of the current members of euro-area.

Knowledge of the mechanism of transmission of monetary policy is also necessary to the CEECs to choose the monetary strategy for the period preceding the accession to the euro-area. The discussion is open, both in the literature and among monetary institutions, on whether a direct inflation targeting or some kind of exchange rate commitment should be preferred: by comparing the mechanism of transmission of policy impulses in different regimes, we can see if a certain strategy made inflation control easier or more difficult, and assess its cost by looking at the impact on the economic activity.

We focused on Poland, Czech Republic, and Slovenia because they are very different for the initial conditions from which they started the transition to functioning market economies and later the integration in the EU, for the relative size and the degree of openness to the international trade, for the policies implemented in due course and even for their historical developments: we think that because of these differences they are, together, representative of the whole group of CEECs. For the euro-area, we took Germany as the reference country.

2. Inflation Control over 1989–2004

2.1. *Transition, Integration, and Accession*

Although Poland, Czech Republic, and Slovenia shared the goals of implementing a functioning market economy and of acceding to the EU, the policies they implemented to reach them were rather different.

The first reason for the difference is in the productive structure at the beginning of the transition. Yugoslavia established a quasi-market system well before 1989, with quite independent firms and relatively hard budget constraints; production was much more centralized and budget constraints much softer in Poland and Czech Republic, and, to make things tougher for these two countries, they also had to suffer the break-up of the council for mutual economic assistance (CMEA), in which they were well integrated (Czech Republic also had to endure the additional trade disruption, caused by the separation from Slovakia). Poland and Czech Republic represent an average position with respect to the general condition of the productive structure of the CEECs at the beginning of the transition. Local differences remain: for example, Hungary implemented some timid reforms before 1989, while the three Baltic countries experienced a much bigger shock, because the disintegration of the Soviet Union resulted in the loss of their largest trade partner. The stronger centralization, the softer budget constraint, and the higher disruption caused by the break. up of the CMEA, and of Czechoslovakia could have *coeteris paribus* caused a more turbulent and longer transition for Poland and Czech Republic, but in reality Slovenia had to face the major shock due to the collapse of Yugoslavia and of the subsequent wars, which deprived the country of its biggest market and had other adverse effects including the loss of revenues and hard currency due to the fall of tourism.

A more precise ranking emerges when the size and the openness to international trade is considered, with Poland being the largest country and Slovenia the smallest one. As far as the monetary policies implemented for inflation stabilization, all the CEECs except Slovenia, began their transition with a strong commitment to a fixed exchange rate. The initial price liberalization, in fact, left the system without a nominal anchor, and, also because of a natural price rigidity with respect to negative corrections, the re-alignment of relative prices took the form of a sudden increase and a steep inflation followed: by opening to the international trade (and by introducing the proper legislation, in order to force the local producers to face a harder budget constraint), the countries in transition had a chance to import a price structure similar to the one of their commercial partners. In fact, being the exchange rate fixed, the domestic producers could not raise their prices too much without losing competitiveness with respect to the foreign ones.

Slovenia followed a different approach to disinflation, targeting the growth of a relatively narrow monetary aggregate (M1); according to Capriolo and Lavrac (2003), anyway the key characteristic of the period was the continuous intervention on the foreign exchange market in order to prevent an excessive real rate appreciation (contrary to the strategy of the Bundesbank and of the Eurosystem, the control of M1 was also enforced with non-market instruments and procedures). Since mainstream optimal currency area literature prescribes a greater incentive to adopt a stable exchange rate to the countries more open to the international trade, the fact that the outcome is actually reversed means that price stabilization rather than trade was the priority in the exchange rate commitment served in Poland and in Czechoslovakia.

Fixing the exchange rate may have contributed to price stabilization, and inflation actually dropped very quickly, but there are certain differences remained with respect to the EU, leading to a strong real exchange rate appreciation. This may have been partially due to the Balassa-Samuelson effect: assuming that the marginal productivity and the wage in the sector of internationally traded goods, is set on the foreign market and that the same wage is transferred to the production of the non-traded goods, then the productivity gap between the two sectors is compensated by a price increase in the less productive domestic sector. This yields a progressive appreciation of the real exchange rate and higher domestic inflation.

The faster growth of productivity, determined by the transfer of technology from the international trade partners and the more efficient allocation of the resources within the economies, compensated part of the pressure on the domestic producers, but the fixed exchange rate arrangements were not sustainable for a long time: Poland switched to a regime of pre-announced crawling peg as early as 1991, later coupling it with an oscillation band which was gradually widened until it was finally abandoned in 2000; Czech Republic resorted to a free float without explicit commitment as early as 1997 under the pressure of a speculative attack, its currency having progressively over-evaluated in real terms over time. Both the countries replaced the monetary anchor by introducing a direct inflation target (Czech Republic starting in 1998, Poland in 1999). The progressive weakening of the exchange rate commitment took place in Hungary and Slovakia too, but in these last two countries, the switch of targets was not complete: Hungary retained a fixed exchange rate commitment, *albeit* with a very wide band, while Slovakia did not implement an explicit inflation targeting because of the high role that administrated prices still had in the consumer prices. Anyway this is not an unanimous attitude: the three Baltic States kept a fixed exchange rate with an extremely tight band, apparently trying to exploit the credibility acquired with it during the first phase of the transition, and Estonia even withstood a moderate speculative attack rather than devaluating. Once again, Slovenia went in the opposite direction: according to Capriolo and Lavraac, a more aggressive exchange rate strategy has been pursued since 2001, with an active, *albeit* informal, crawling peg.

2.2. *Exchange Rate and Direct Inflation Targeting*

The fixed exchange rate played a key role in the stabilization phase; nowadays, it is still the reference of monetary policy in the Baltic States, and it is also present in Hungary. Since, the exchange rate commitment is also part of the Maastricht criteria, where two years of participation to the exchange rate mechanism (ERM) 2 is required, the decision of Poland, Czech Republic, and Slovakia, to completely abandon it may indeed seem to be a paradox.

Surely, a small country cannot ignore the effect of the exchange rate fluctuations on the stability of financial institutions and the economy as a whole, but a commitment to a fixed parity is a much stronger policy. Whether, macroeconomic stability and inflation control are helped or hindered by this depends on the credibility of the commitment itself.

There can be little doubt that an exchange rate commitment is a risky policy: the integration of the financial markets allows the mobilization of large amounts of funds, and no central bank can muster enough foreign exchange reserves to resist a sustained attack, nor indeed may desire to do it, because the mere defence of the currency could require a substantial rise in the domestic interest rates, for which the economy may be even more detrimental than the devaluation that it was trying to prevent. Against an exchange rate commitment, stands the evidence of speculative attacks in the past: consider, for example, the break up of the ERM 1, when even the French currency suffered some pressure, although the fiscal, financial, and macroeconomic situation of the country was not worse than it was in Germany. The same lesson can indeed be derived by looking at the

Hungarian experience: when the two objectives were in conflict, as in June 2003, one had to be abandoned, and a devaluation took place.

Since the accession to the EU also required the complete adoption of the *acquis communautaire*, thereby including the elimination of capital controls, the apparent paradox of several CEECs switching to a free float and inflation targeting, for the period preceding the adoption of the euro, is then a strategy that may protect their currencies from unrealistic parities and speculative attacks.

On the other hand, a direct inflation targeting can only be successful if the monetary authority has earned itself a solid reputation, and this primarily requires the independence from any political interference. Since the CEECs only acquired a formal central bank independence in the last few years, as a part of the process of adoption of the *acquis communautaire*, their credibility may still be weak. Central bank independence is even more a concern, when the definition is extended beyond the mere legal requirement: in most of the cases for example, the credibility was seriously weakened because the fiscal authority had the opportunity to partially finance its expenditures through the central bank; political pressures, as experienced in the Czech Republic and in Poland, may undermine the credibility of the central bank even when they are resisted, because they may erode the consensus in the country. Several indices of central bank independence, were proposed in the literature: according to Maliszewski (2000), and to the extensive survey in Dvorsky (2000), the CEECs were lagging behind with respect to their Western counterparts, Poland and the Czech Republic faring better than Slovenia. Indeed, it is exactly because the credibility of the monetary authorities is lower in the CEECs that Amato and Gerlach (2002) proposed to supplement the inflation targeting with a mild monetary commitment, this would increase the information in the market.

It is then important to ascertain if the exchange rate commitment is a risk worth taking, not only for the countries on the way to EMU that are still more than two years away from the formal discussion of their convergence according to the Maastricht criteria, but also for other emerging economies in the world.

2.3. *A Summary of the Results of Previous Analyses*

Several empirical analyses on inflation control in CEECs appeared in the recent years. Iacone and Orsi (2004) surveyed many applied works: these varied for the country that was analyzed, for the period considered and for the methodology adopted, making comparison rather difficult.

Interpretation of the results was often dubious because either the variables of interest were replaced by first or second differences, or the intermediate steps linking the monetary instrument to the target were omitted, thus obscuring any potential evidence about the channel through which inflation may be controlled. Reliability was also hampered by the fact that in many cases no stability analysis was provided, despite the potentially disruptive effects of the transition on the estimates, nor any attempt was made to model the slow formation of a market economy despite this being the main feature of the period.

It seems nevertheless fair to conclude that the exchange rate is very important for the stabilization of inflation, an appreciation of the real exchange rate reducing the growth rate of prices. The empirical evidence supporting the existence of a standard interest rate channel was much more controversial, the results depending largely on the econometric approach and model adopted by the researcher and on the sampling period considered.

2.4. *A Structural Model for Monetary Policy*

It is possible that the weak support provided by these analyses is at least partially due to the methodology adopted: VAR models have the advantage of requiring a minimal structural specification, but, in return, they need the estimation of many parameters. Considering the short length of the sampling size and the potential extreme instability of the parameters due to the transition, large standard errors and non-significant estimates seem to be unavoidable consequences.

A structural model can provide a more parsimonious approach: we followed Rude-bush and Svensson (1999) and considered the model

$$\begin{cases} \pi_t = \alpha_0 + \alpha_{\pi 1}\pi_{t-1} + \alpha_{\pi 2}\pi_{t-2} + \alpha_{\pi 3}\pi_{t-3} & \\ \qquad + \alpha_{\pi 4}\pi_{t-4} + \alpha_y y_{t-1} + \varepsilon_{\pi,t} & \text{PC} \\ y_t = \beta_0 + \beta_y y_{t-1} - \beta & \text{AD} \end{cases}$$

where y_t is a measure of the economic activity, π_t is the annualized inflation within the quarter, it is a short-term interest rate, $\bar{\pi} = \frac{1}{4}\sum_{j=0}^{3} \pi_{t-j}$ is the average inflation rate in the last four quarters (i.e., the inflation over the last year), and $\bar{i} = \frac{1}{4}\sum_{j=0}^{3} i_{t-j}$ is the average interest rate over the same period.

The equations are referred to as Phillips Curve (PC) and aggregate demand (AD), respectively because the first one can be given a structural interpretation as a PC while the second one may describe the transmission of monetary policy on the economic activity quite like in an AD function.

With this specification, agents extrapolate from the past inflation to formulate the current period price level. The mechanism of transmission of monetary policy to inflation is indirect: an expansionary monetary policy takes place as a decrease in the real rate $\bar{i}_{t-1} - \bar{\pi}_{t-1}$, which boosts the economic activity in the AD equation; in the second moment, the pressure on the demand enhances inflation, as illustrated in the PC equation. The instrument and the mechanism of transmission of monetary policy are then consistent with the operating procedures of both the FED and of the Bundesbank, and later of the ECB.

Rudebush and Svensson (1999) allowed for an additional term $\beta_{y2}y_{t-2}$ in the AD equation, but we found that, possibly also due to the smaller dimension of our samples and the high collinearity with $\beta_{y2}y_{t-2}$, the contribution of the two different effects in general could not be distinguished. We, then considered the proposed AD equation sufficient, also because one lag proved to be enough to have no serial correlation in the residuals. Rudebush and Svensson also specified the model without intercepts α_0, β_0, having formulated for mean-corrected data, but we preferred to allow for the constant and test, for its possible exclusion explicitly.

Rudebush and Svensson (2002) remarked that $\sum_{j=1}^{4} \alpha_{\pi j} = 1$ can be expected as it corresponds to a vertical PC in the long run. They also explicitly discussed the fact that the backward looking formulation of the PC and AD equations imply adaptive rather than rationale expectations, but they argued that this may well be the case especially in a phase of monetary and inflationary instability; as an alternative, they suggested that the given specification may simply be considered as reduced form equations.

Notice that even if the constraint $\sum_{j=1}^{4} \alpha_{\pi j} = 1$ is applied, the model still does not necessarily impose a unit root to inflation, because the dynamics of y_t has to be taken into account too: consider for example $\alpha_{\pi 1} = 1$, $\alpha_{\pi 2} = \alpha_{\pi 3} = \alpha_{\pi 4} = 0$, $\beta_y = 0$, and the monetary policy rule $\bar{i}_t = \bar{\pi}_t + \pi_t$; the PC equation then becomes $\beta_r \in y, t+ \in \pi, t$, thus, inducing a stationary behavior. It is indeed, precisely by steering the output gap using the interest rate that the stabilization of inflation is ultimately possible.

This model may well be applied to the transition economies too, especially considering that the accession of these countries to the EU requires them to adopt a monetary policy setting that is institutionally compatible with the one of the ECB. In the case of a small open economy, though, two other factors should be taken into account: the demand of domestic products generated by the trade partners and the effect of the international competition on local prices.

Following Svensson (2000), we then augmented the model by introducing the level of the economic activity in the international partners wt and the percent real exchange rate depreciation $(q_t - q_{t-1})$, where $q_t = p_t^* + e_t - p_t$ is the logarithm of the real exchange rate and p_t, p_t^* and e_t are the logarithm of the local and foreign level of prices and of the nominal exchange rate, respectively. Both the variables are added to the AD equation, and the real exchange rate is added to the PC equation as well. For the AD, the assumption is that phases of high economic activity abroad result in higher demand of locally produced goods, while a too high level of prices (in real terms) causes a shift of the domestic and foreign demand towards the external producers. The real exchange rate entered the PC equation, because the international competition imposes some price discipline to the local producers: for given foreign prices, a nominal exchange rate appreciation makes foreign goods cheaper in local currency, thus, forcing the domestic producers to lower their prices to keep up with the external competitors, while for given nominal exchange rate higher foreign prices give the domestic producers to set higher prices and stay in the market; we also refer to Svensson (2000), where he explains how an increase in foreign prices in local currency, either due to the move of the foreign price itself or to the exchange rate, results in higher cost of inputs and then feeds back in higher prices of output.

Maintaining the autoregressive structure of Rudebush and Svensson, we describe the economy with the two equations model

$$\begin{aligned}
\pi_t &= \alpha_0 + \alpha_{\pi 1}\pi_{t-1} + \alpha_{\pi 2}\pi_{t-2} + \alpha_{\pi 3}\pi_{t-3} \\
&\quad + \alpha_{\pi 4}\pi_{t-4} + \alpha_y y_{t-1} - \alpha_q(q_{t-1} - q_{t-5}) + \varepsilon_{\pi,t} \qquad \text{PC} \\
y_t &= \beta_0 + \beta_y y_{t-1} - \beta_w w_{t-1} - \beta_q(q_{t-1} - q_{t-5}) + \varepsilon_{y,t} \ \text{AD}
\end{aligned}$$

where we actually considered relevant for the exchange rate the depreciation over the whole year.

In the original model, Svensson considered the real exchange rate rather than the percent depreciation: in our model specification, we preferred the latter because as we saw the real exchange rates of most of the CEECs appreciated since the beginning of the transition without resulting in a dramatic decline of the economic activity.

Moreover, we kept the AD/PC structure as a general reference, but given that the specification is somewhat arbitrary, we tried to adapt it to the economies considered. Our sample starts from 1990, and to avoid the risk of model instability in the very first part of the sample, the data were used only to compute the output gap or as lags. In some cases, we did not find adequate data so the sample period was even shorter. Data are monthly, but in order to reduce the volatility, we used quarterly averages, adjusting the frequency properly. We took the CPI to compute inflation and the real exchange rate, and the seasonally adjusted industrial production to derive a measure of economic activity (when only the nonseasonally adjusted series was available, we run a preliminary step using the X-12 filter in EViews to remove seasonality); for the nominal interest rate we used a three-months inter-bank rate or the Treasury bills rate of the same maturity. More details on the data used are at the end of this paper. We always considered Germany as the international reference, even if in the first part of the sample period some countries pegged the exchange rate to the US$: the EU in fact, constitutes by far the biggest trade partner for the CEECs.

There is a strong seasonal component in the quarterly inflation, which is likely to shift the weights towards the fourth lag in the PC equation; the choice of the year appreciation of the real exchange rate was then also convenient to remove the seasonal effect.

As a measure of the economic activity we took the output gap, defined the difference between the observed and the potential output computed using the Hodrik Prescott (HP) filter on the quarterly average of the production index. Even if the HP gap is usually considered an inexact measure of the economic cycle, due to its purely numerical definition, we still think that it is at least informative of the effective output gap. Plots of the inflation in the four countries are in Fig. 1 (figures are all collected at the end of

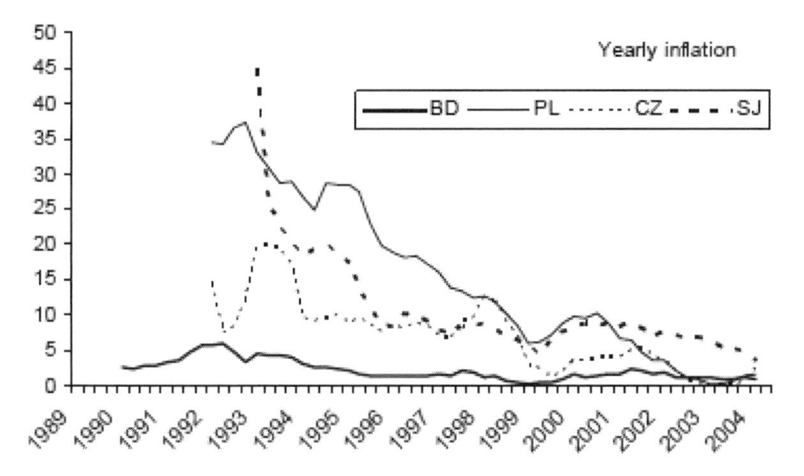

Figure 1. Yearly inflation, Germany, Poland, Czech Republic, and Slovenia.

the work). The inflation used for this graph was defined as the growth rate of prices on the previous year: we presented it *albeit*, we used a different measure in our empirical study because it is very common in the descriptive analysis in the literature. Clearly, though, averaging the inflation in this way induces spurious autocorrelation in the data, so in the empirical analysis we measure inflation as the annualized growth rate of prices over each period. We plotted this and the output gap for Germany in Fig. 2; in Figs. 3–5, we plot the same data and the yearly real exchange rate depreciation for Poland, Czech Republic, and Slovenia, respectively. From these figures, it

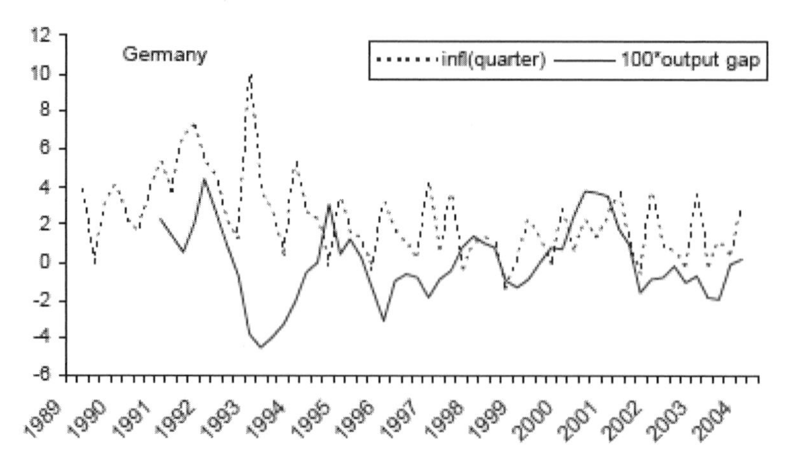

Figure 2. Quarterly inflation and output gap, Germany.

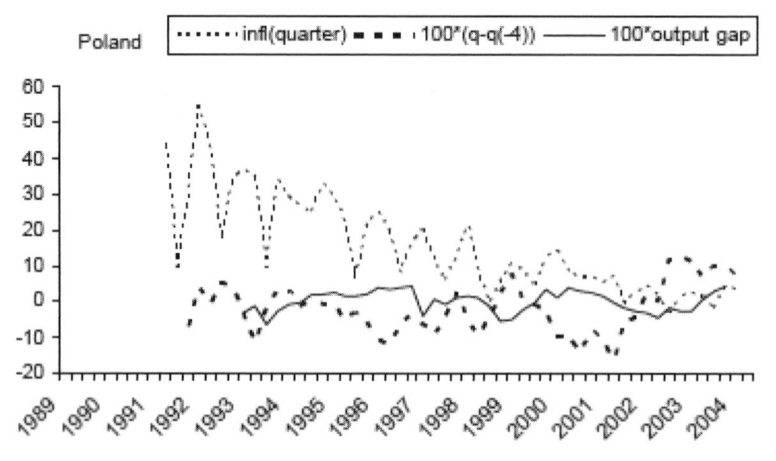

Figure 3. Quarterly inflation, output gap, real exchange rate depreciation, Poland.

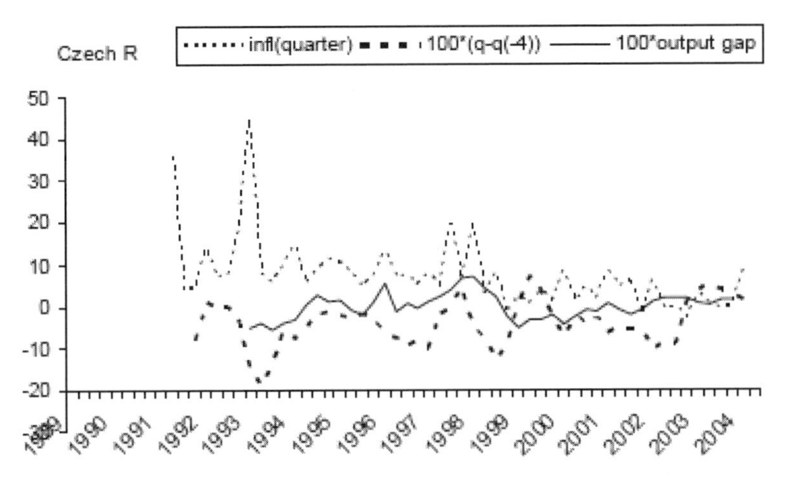

Figure 4. Quarterly inflation, output gap, real exchange rate depreciation, Czech Republic.

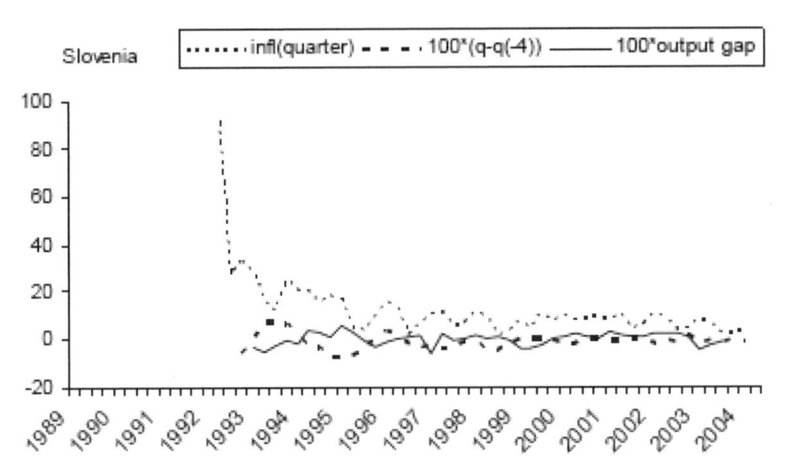

Figure 5. Quarterly inflation, output gap, real exchange rate depreciation, Slovenia.

is already possible to notice that, while for Germany, the dynamics of inflation is dictated by the output gap, in the three new EU members inflation is largely dictated by the real exchange rate dynamics.

Estimation was carried out using OLS equation by equation, as in Rudebush and Svensson (1999). We did not introduce any other modification to the AD equation, but we re-wrote the PC as

$$\Delta\pi_t = \alpha_0 + \rho_0\pi_{t-1} + \rho_1\Delta\pi_{t-1} + \rho_2\Delta\pi_{t-2} + \rho_3\Delta\pi_{t-3} + \rho_y y_{t-1} + \alpha_q(q_{t-1} - q_{t-5}) + \varepsilon_{\pi,t} \qquad \text{PC}$$

(the term $q_{t-1} - q_{t-5}$ does not appear in the equation for Germany). This is a re-parametrization of the original equation, with

$$\rho_0 = (\alpha_{\pi 1} + \alpha_{\pi 2} + \alpha_{\pi 3} + \alpha_{\pi 4} - 1)$$

$$\rho_1 = -(\alpha_{\pi 2} + \alpha_{\pi 3} + \alpha_{\pi 4}),\ \rho_2 = -(\alpha_{\pi 3} + \alpha_{\pi 4}),\ \rho_3 = -\alpha_{\pi 4}$$

Being simply a re-parametrization, the residual sum of squares that has to be minimized is the same, so the OLS estimates are not different than the estimates that one would obtain in the original model in levels. The advantage of this specification is that long and short dynamics are explicitly separated, $\sum_{j=1}^{4} \alpha_{\pi j} = 1$ corresponding to $p_0 = 0$, and that, more importantly, the multi-collinearity due to the terms π_{t-1} to π_{t-4} is much less severe.

Correct specification was primarily checked by looking at the tests for normality (Jarque and Bera), autocorrelation (up to the fourth lag: Breusch .Godfrey LM) and heteroskedasticity (white without cross terms) in the residuals: we referred to these tests using N for the normality, AC for the autocorrelation, and H for the heteroskedasticity. If the model appeared to be correctly specified, we then analyzed the stability of the parameters over time. If we did not reject the hypothesis of stability either, we moved to test specific restrictions on the parameters, and to estimate a more efficient version of the model. In all the tests, we used the conventional 5% size. In the tables in the Appendix, we presented both the P-values associated to the tests and a summary of the equation estimated when the restriction was imposed.

We considered the normality test first, because we used its result to choose between the small or large sample version of the other tests. Since the small and large sample statistics are asymptotically equivalent in large samples, the asymptotic interpretation of the results is not affected. Admittedly, due to the asymptotic nature of the Jarque and Bera test, and to the small sample lower order bias in the estimation of autoregressive coefficients, the reader may be cautious in the interpretation of the results, especially, when only portions of the dataset are used to estimate parameters (like in the Chow stability test or in the estimates on sub-samples): notice anyway that this is a problem that depends on the dimension of the sample; so, indeed the same caveat applies to all the empirical analyzes present in the literature. The reader may then want to consider the estimates and the statistics computed on the estimated parameters more as general indications in small samples (especially, if the sub-sample 1998–2004 only is used), but we think that despite these drawbacks, the estimates and the tests do still provide some information on the structure that we intend to analyze, *albeit* an approximate one.

Autocorrelation in the residuals is a serious mis-specification since it results in inconsistent estimates given that both the equations include a lagged dependent variable. Heteroskedasticity is a minor problem, because estimation is simply inefficient (compared to the GLS estimator) but not inconsistent. Yet, since the standard errors indicated by the regression are not correct, evidence of heteroskedasticity requires at least a different estimation of the variance-co-variance matrix of the estimates.

Due to the peculiarity of the transition, we think it is particularly important to control the parameter stability over time, and possibly to model it when apparent. We used as preliminary diagnostic the CUSUM squared test, but also considered as a particular breakpoint in 1998, because of the opening of the negotiation to access the EU. Clearly, this is just a broad indication: one could also argue that by that time the negotiation opened the transition to a market economy was already nearly complete, and the issue was convergence to the EU standards. Yet, that period may still be informative about a potential break even if it was not located exactly there. We also considered country-specific breakpoints in the PC equations to analyze the effect on the inflation dynamics of monetary regime changes: for Germany, we allowed a break in 1999Q1, when the control of monetary policy was transferred from the Bundesbank to the ECB, because it may have been perceived as a weakening of the anti-inflationary stance, inducing higher inflationary expectations; for Poland, we considered a break in 2000Q2, when the exchange rate commitment was formally abandoned; for Slovenia, the change is in 2001Q1, when the approach towards the real exchange rate became effective. We tested the presence of the break with a standard Chow test, unless the sample after the break was very little, in this case we used the Forecast test, which is exactly designed to test for breaks when only a little data are located after the break.

If the model was correctly specified and stable, we improved the efficiency of the estimates by removing the variables whose coefficients were estimated with a sign opposite to the one prescribed in the economic theory (we assumed that $\alpha_y, \alpha_q, \beta_y, \beta_w, \beta_q, \beta_r$ are not negative), and by imposing other restrictions (including $\sum_{j=1}^{4} \alpha_{\pi j} = 1$): we tested these with a Wald statistic, and referred to these tests as W in the tables.

Of course, given the small dimension of the samples and the potential instability of the models due to the transition, a certain caution should still be exerted in the interpretation of the results. But, we think that even in this case our analysis is at least able to get a general, broad picture of the situation: the lack of precision that may be attributed to the small dimension of the sample, for example, is already taken into account in the standard error. If, under normality, a test statistic rejects the null hypothesis despite the large variability due to the small sample, then we would still regard the estimate as informative, although may be only on the sign.

3. The Empirical Evidence

3.1. *Germany*

As a benchmark case, we first estimated the model for Germany over the period 1991Q1–2004Q1. The choice of the sample is two-fold: first, only data after the German re-unification were considered, leaving many of the problems related to that particular transition out of the discussion; second, this time span is roughly the same one covered by the data of the CEECs in most of the applied analyzes, making the results comparable in this sense. We called this a "benchmark" model because it is widely assumed that this transmission mechanism for monetary policy was at work in Germany in the years we are interested in. Furthermore, Germany is geographically and economically close to the CEECs, accounting for a large share of their international trade. Finally, it is an interesting reference because Germany is the biggest country in the euro-area: if the monetary transmission in the CEECs works in a very different way, their integration in the euro may adversely affect the dynamics of inflation.

The estimated model for Germany was

$$\begin{cases} \hat{y}_t = \underset{(0.005)}{0.011} + \underset{(0.084)}{0.769} y_{t-1} - \underset{(0.00163)}{0.0038} y_{t-1} \\ \widehat{\Delta\pi_t} = -\underset{(0.10)}{1.02}\Delta\pi_{t-1} - \underset{(0.12)}{0.86}\Delta\pi_{t-2} - \underset{(0.10)}{0.71}\Delta\pi_{t-3} + \underset{(12.18)}{17.62} y_{t-2} \end{cases}$$

over the period $1991Q2$–$2004Q1$ for the AD, and $1993Q1$–$2004Q1$ for the PC.

According to the diagnostic tests, the AD equation was correctly specified and stable over the whole sample; the CUSUMsq hit the boundary in the period 1993–1996, but we found no evidence of break with the Chow test (further details about all the tests and a summary of the estimates are in the tables at the end).

The inflation dynamics, on the other hand, was not stable when the whole sample was taken into account. The instability, though was clearly restricted to the first part of the sample, and could indeed be removed dropping the years 1991 and 1992. It is interesting to observe that it is exactly the period of the recovery of the inflationary shock due to the German re-unification, a phenomenon that was usually regarded as temporary and exceptional. Diagnostic tests confirmed that the equation estimated using data from 1993 onwards was correctly specified and stable; notice anyway, the distribution of the residuals did not appear to be normal, so the interpretation of the other tests can only be justified asymptotically. Point estimates of ρ_1, ρ_2, and ρ_3 were all very close to -1, suggesting a very high value of $\alpha_{4\pi}$, or a strong seasonal component in inflation; we also found that lagged values of $yt - 1$ had a more significant effect, *albeit* even in this case the P-value of the test H_0: $\{\alpha y = 0\}$ vs. H_0: $\{\alpha_y > 0\}$ was between 5% and 10%.

Considering the model as a whole, anyway, we find these estimates satisfactory, although not as much as those presented by Rudebush and Svensson for the United States: we conjecture that the lower precision of the estimates for Germany depended on the smaller number of observations available.

3.2. *Poland*

Opposite to Germany, we consider Poland (and the other CEECs later) as a country small enough to be affected by the international trade, so we augmented the model with the real exchange rate depreciation with respect to Germany and with the German output gap. *Albeit* our dataset included 1993 too, we found that both the equations suffered from residual autocorrelation, resulting in inconsistent estimates. We interpreted this as evidence of instability, because when 1994 was taken as a starting point, the residual autocorrelation was largely removed and the other diagnostic tests were broadly compatible with a stable, correctly specified model. The estimated model was

$$\begin{cases} \hat{y}_t = \underset{(0.008)}{0.020} + \underset{(0.142)}{0.467} y_{t-1} - \underset{(0.001)}{0.0027} y_{t-1} \\ \widehat{\Delta\pi_t} = -\underset{(0.12)}{0.64} \Delta\pi_{t-1} - \underset{(0.101)}{0.67} \Delta\pi_{t-2} - \underset{(0.10)}{0.51} \Delta\pi_{t-3} + \underset{(7.88)}{21.46} (q_{t-1} - q_{t-5}) \end{cases}$$

over the period $1994Q1$–$2004Q1$ for the AD, and $1994Q1$–$2004Q1$ for the PC.

The variables referred to the trade partners were then not significant in the AD equation: we found a certain effect at least for the real exchange rate depreciation, and with a sign consistent with the macroeconomic theory, but the realization of the t statistic was still rather below the critical value. The absence of the two variables referring to the international trade could be interpreted as a sign that Poland is still large enough to be more sensible to the internal rather than to the international developments, but notice that the situation was reversed for the PC equation, where only the external conditions had a significant effect on the inflationary dynamics (again we found that the effect of the omitted explanatory variable, y_{t-1} in this case, had the sign predicted by the macroeconomic theory, but it failed to be significantly different than 0).

We concluded performing a Forecast test for a break in $2000Q2$ in the PC equation, rejecting it (P-value 0.65). Since at that point, the Polish national bank abandoned the exchange rate commitment and left the direct inflation targeting to stand alone, it is fair to conclude that the change of monetary target did not result in a destabilization of the inflation dynamics. Since the target for monetary policy is defined mainly because it is assumed that in that way the expected inflation can be favorably influenced, we prefer not to give a structural interpretation of a backward looking formulation of the inflation equation, but in this case we can at least conclude that the change of target did not result in a worsening of the expectation of inflation. Thus, it is likely that by 2000, the Polish central bank acquired sufficient credibility to move away from a commitment that would have exposed it to pressure from market speculation.

Although we summarized the diagnostic tests, concluding that the instability that may be associated to the transition to a market economy can be largely confined to the period before 1994, we also estimated a more restrictive specification, in which the sample for the AD equation is from 1995 onwards, and for the PC from 1998 onwards. We considered this second specification, because the residuals were not normally distributed, and an asymptotic interpretation of the outcome of the autocorrelation test for the AD equation was particularly dubious because the test statistic reached approximately, the limit critical value, the P-value being slightly lower than 0.05 (it was 0.046). As far as the PC equation, the Chow test did not identify any break in 1998Q1 but the CUSUMsq statistic did indeed signal a potential break in this year, *albeit* only mildly. The estimates in the more restrictive model are

$$\begin{cases} \hat{y}_t = \underset{(0.009)}{0.021} + \underset{(0.154)}{0.454} y_{t-1} - \underset{(0.001)}{0.0029} y_{t-1} \\ \widehat{\Delta\pi_t} = -\underset{(0.19)}{0.57}\Delta\pi_{t-1} - \underset{(0.13)}{0.76}\Delta\pi_{t-2} - \underset{(0.20)}{0.47}\Delta\pi_{t-3} + \underset{(7.23)}{16.09}(q_{t-1} - q_{t-5}) \end{cases}$$

over the period 1995$Q$1–2004$Q$1 for the AD, and 1998$Q$1–2004$Q$1 for the PC, very similar then to the estimates on the longer sample. For the PC equation, we also present the alternative specification

$$\widehat{\Delta\pi_t} = -\underset{(0.20)}{0.66}\Delta\pi_{t-1} - \underset{(0.13)}{0.78}\Delta\pi_{t-2} - \underset{(0.20)}{0.59}\Delta\pi_{t-3} + \underset{(29.00)}{39.24}\Delta\pi_{t-1} + \underset{(7.46)}{19.22}(q_{t-1} - q_{t-5})$$

estimated over the sample 1998$Q$1–2004$Q$1: the lagged output gap still appeared in the determinants of the inflation dynamics, it had the correct sign and it was also marginally significant with the 10% critical value and a one-sided t test: a possible interpretation of this result could be that by 1998, the PC equation evolved even closer to the standard textbook specification, but we think that a longer dataset is needed to address this particular issue.

In any case, considering the whole model, we confidently conclude that the Polish monetary authority was able to control inflation during these years, but we think it mainly worked through the exchange rate management, the evidence of the closed economy transmission mechanism, in fact, being dubious.

3.3. *Czech Republic*

Czech Republic originated in 1993, from the split of the former Czechoslovakia. Data availability is then slightly reduced, and it is also possible that the division of the country acted as a source of instability additional to the one generated to the transition. Although it is not possible to distinguish between these two causes, the evidence of instability of the macroeconomic structure was quite clear.

The sample for the estimation of the AD equation started in 1994$Q$4, but we found that a break took place around 1999, according to the CUSUMsq statistic. This is approximately also the point for which the Chow test is to be computed, and we found strong evidence in favor of it (the P-value associated to it was 0.01). Comparing the estimates in the two sub-samples, the most relevant differences were in the fact that the coefficient of the real rate, that signals the transmission of the monetary policy impulse from the monetary authority to the economy, and the real exchange rate depreciation, were only significant and with the correct sign in the second part of the sample (the economic activity in Germany remained insignificant).

It is possible that instability affected the PC equation too, *albeit* in this case the evidence was less compelling. Using the whole sample that started in 1994, the CUSUMsq statistic did not indicate the presence of any break, but according to the Chow test a discontinuity took place in 1998. The estimated coefficient of the economic activity was largely insignificant, despite having the same sign predicted by the economic theory, while we found that the real exchange rate appreciation had a strongly significant anti-inflationary effect. The estimated model was

$$\begin{cases} \hat{y}_t = \underset{(0.006)}{0.010} + \underset{(0.116)}{0.620} y_{t-1} - \underset{(0.001)}{0.03} y_{t-1} - \underset{(0.050)}{0.083} (q_{t-1} - q_{t-5}) \\ \widehat{\Delta\pi_t} = -\underset{(0.23)}{0.76} \Delta\pi_{t-1} - \underset{(0.25)}{0.21} \Delta\pi_{t-2} - \underset{(0.18)}{0.31} \Delta\pi_{t-3} + \underset{(11.97)}{32.55} (q_{t-1} - q_{t-5}) \end{cases}$$

over the period $1998Q1$–$2004Q1$ for both the AD and the PC, while if we relied on the CUSUMsq statistic rather than on the Chow one for the PC, the PC equation estimated over the period $1994Q1$–$2004Q1$ was

$$\widehat{\Delta\pi_t} = -\underset{(0.15)}{0.84} \Delta\pi_{t-1} - \underset{(0.17)}{0.33} \Delta\pi_{t-2} - \underset{(0.09)}{0.34} \Delta\pi_{t-3} + \underset{(9.186)}{25.338} (q_{t-1} - q_{t-5})$$

Comparing these estimates, notice that if a break did indeed take place, then the consequence was that the real exchange rate appreciation had a stronger effect against inflation in the last part of the sample, that was exactly when Czech central bank used inflation rather than exchange rate targeting. As in the case of Poland, we then found no evidence that the switch of monetary policy target resulted in a worsening of the inflationary dynamics. On the contrary, if a change in the dynamics of the growth rate of prices took place at all, when inflation targeting was introduced, it was in favor of an easier control of it.

3.4. *Slovenia*

As we did for Czech Republic, in the case of Slovenia too we only considered the part of the sample corresponding to the existence of an independent state.

The data available for the estimation started from 1994Q1 for the AD equation and in 1993Q2 for the PC. We found anyway, evidence of model mis-specification when the first year is included, resulting in residual autocorrelation and inconsistent estimates: it is also fair to suspect, on the basis of the CUSUMsq statistic, that a break took place in the first part of the sample. We then removed 1993 from the dataset, and estimated the model from 1994 onwards (sample period 1994$Q$1–2003$Q$3 for the AD, 1994$Q$1–2003$Q$4 for the PC), obtaining

$$\begin{cases} \hat{y}_t = \underset{(0.167)}{0.0226} + \underset{(0.234)}{0.343} w_{t-1} \\ \widehat{\Delta\pi_t} = -\underset{(0.14)}{0.60}\Delta\pi_{t-1} - \underset{(0.12)}{0.66}\Delta\pi_{t-2} - \underset{(0.12)}{0.45}\Delta\pi_{t-3} + \underset{(15.60)}{52.64}(q_{t-1} - q_{t-5}) \end{cases}$$

The diagnostic tests confirmed that both the equations were correctly specified and stable in the period considered, *albeit*, we acknowledge that the estimates in the AD equation are only significant using a 10% threshold. We are inclined to consider them anyway, suspecting that a more clear link will emerge in the future, with a longer sample and an even higher integration in the European economy. Notice, on the other hand, the absence of the real interest rate, whose estimated coefficient even failed to have the sign prescribed in the economic theory: this supported the argument of Capriolo and Lavraac that, until recently, the interference of the central bank on the financial markets made the interest rates not informative about the monetary stance.

The estimated PC is qualitatively similar to the Polish and Czech one, with inflation only driven by the exchange rate dynamics but not by the output gap: indeed, $\hat{\beta}_y$ even resulted as negative, so in the final specification we excluded y_{t-1} from the explanatory variables altogether. In order to see, if the more active exchange rate policy resulted in an easier inflation control and then in a larger β_q, we also tested (with a Forecast test) for a break in 2001, but we did not reject the hypothesis of stability.

3.5. *Some Concluding Remarks*

Poland, Czech Republic, and Slovenia are small economies that in a few years re-shaped their productive and distributive structures, opened to the international trade, established a functioning market economy and joined the EU. Despite these common features, they differ quite remarkably for the time and the condition, in which they started the transition to the market economy, for the policies they implemented since 1989, and for the size and the relative importance of the trade with the rest of the EU. It is then fair to expect that both the similarities and the differences emerge, and this is indeed the case.

The most relevant common feature is the presence of a certain instability in the first part of the sample. To appreciate the importance of the transition, we observed that, according to a study of Coricelli and Jasbec (2004), the characteristics that may be associated to the initial conditions were the main driving forces of the real exchange rate dynamics. The variables that were more directly related to the economic notions of supply and demand, only acquired importance after a period of approximately five years. We found that the macroeconomic model structure begun to show a stable pattern only from 1994 onward, confirming the findings of Coricelli and Jasbec. Even taking 1994 as a starting value, anyway, some elements of instability remained, especially with respect to Czech Republic.

The instability of the estimated macroeconomic relations is also informative about the speed of the transition. In this respect, we found that in Slovenia the adverse effect of the Yugoslavian civil war canceled the advantage due to the self-management reform; Czech Republic, which had to face not only the decentralization and the break-up of CMEA but even the split of Czechoslovakia, fared worse.

The relation between the variables got closer to the standard textbook description of a market supply and demand structure in the last few years. Yet, even in that part of the sample evidence of a transmission mechanism of monetary policy based on the interest rates was only present for Poland, and that too turned out to be very weak. Although it is not surprising that the exchange rate has a prominent role in inflation control in small open economies, our findings cannot be explained simply by this argument, because we explicitly include the exchange rate in both the equations of the model, then taking it into account (and holding it fixed in the interpretation of the results) in the multiple regression framework. One has rather to conclude that the interest rate channel of transmission of monetary policy was weak or obstructed, as indeed explicitly argued for Slovenia. This also means that, although these economies were acknowledged as functioning market economies, they may still be lagging behind; nonetheless, since a good development took place during the accession period, it is fair to expect that participation to the EU will foster integration. Meanwhile, their presence should not hamper inflation control too much, because the exchange rate is clearly a valid instrument for that purpose: this means that in a widened euro-zone, the Eurosystem may control inflation in the CEECs by controlling it in the core area.

Although they are all small open economies, a pattern can be envisaged in terms of dimension and openness to international trade, and this is reflected in the estimated models: Poland, is the only country in which the international factors did not affect the AD. Another way to appreciate the importance of the international openness is to look at the estimated effect of a 1% appreciation of the exchange rate: the impact on the inflation dynamics is clearly ranked according to the relative dimension of trade.

Stability analysis is also important because it allows us to understand if the switch in Poland and in Czech Republic from exchange rate to inflation targeting affected the price setting formation. We did not find any evidence of it, nor we found that a more aggressive exchange rate policy in Slovenia eased the costs of disinflation from 2001 onwards. The instability that we found for Czech Republic seems to suggest an evolution of the macroeconomic structure towards a market economy, and should then have more to do with EU accession than with inflation targeting. One main implication that we perceive is that the exchange rate commitment is either irrelevant or, more likely, replaceable with inflation targeting as a nominal anchor. We think that this conclusion is of particular interest to help choosing the monetary target for the period preceding the last two years before fixing the exchange rate in the ERM2: inflation targeting should be preferred and ERM2 membership should only be limited to the two mandatory years. We also think that this lesson can be extended to other emerging market economies, and that inflation targeting, rather than explicit exchange rate commitment, should be preferred, especially if strict capital controls are not in place.

References

Amato, JD and S Gerlach (2002). Inflation targeting in emerging market and transition economies: lessons after a decade. *European Economic Review* **46**, 781–790.

Capriolo, G and V Lavrac (2003). Monetary and exchange rate policy in Slovenia. *Eastward Enlargement of the Euro-Zone Working Papers*, 17G. Berlin: Free University Berlin, Jean Monnet Centre of Excellence, 1–35.

Coricelli, F and B Jasbec (2004). Real exchange rate dynamics in transition economies. *Structural Change and Economic Dynamics* **15**, 83–100.

Dvorsky, S (2000). Measuring central bank independence in selected transition countries and the disinflation process. *Bank of Finland*, Bofit Discussion Paper 13. Helsinki: Bank of Finland, 1–14.

Iacone, F and R Orsi (2004). Exchange rate regimes and monetary policy strategies for accession countries. In *Fiscal, Monetary and Exchange Rate Issues of the Eurozone Enlargement*, K Zukrowska, R Orsi and V Lavrac (eds.), pp. 89–136. Warsaw: Warsaw School of Economics.

Maliszewski, WS (2000). Central bank independence in transition economies. *Economics of Transition* **8**, 749–789.

Rudebush, GD and LEO Svensson (1999). Policy rules for inflation targeting. In *Monetary Policy Rules*, John B Taylor (ed.), pp. 203–246. Chicago: Chicago University Press.
Rudebush, GD and LEO Svensson (2002). Eurosystem monetary targeting: lessons from U.S. data. *European Economic Review* **46**, 417–442.
Svensson, LEO (2000). Open-economy inflation targeting. *Journal of International Economics* **50**, 155–183.

Appendix A
Summary of the Results

Note to the Tables: for each equation, N indicates the normality test, AC is the LM test of no autoregressive structure in the residuals and H is the test for the heteroskedasticity; W indicates the test of those restrictions that reduce the corresponding equation as specified in general to the one specifically used. For each test, we report the P-value: the small sample version of the test is taken, unless normality is rejected. When the estimated equation is restricted, the test statistics N, AC, H, Chow, W are calculated in the unrestricted model.

Table 1a. Germany.

AD [sample: 91Q2–04Q1]

$\hat{y}_t = 0.011 + 0.769_{y_t-1} - 0.038r_{t-1}$

[N: 0.38] [AC: 0.20] [H: 0.17]

PC [sample: 93Q1–04Q1]

$\hat{\Delta}\pi_t = -1.02\Delta\pi_{t-1} - 0.86\Delta\pi_{t-2} - 0.71\Delta\pi_{t-3} + 17.62y_{t-2}$

[N: 0.00] [AC: 0.24] [H: 0.10] [Chow: 0.18] [W: 0.44]

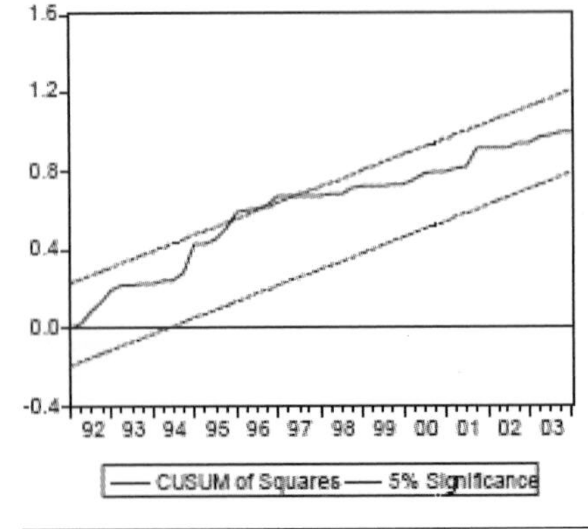

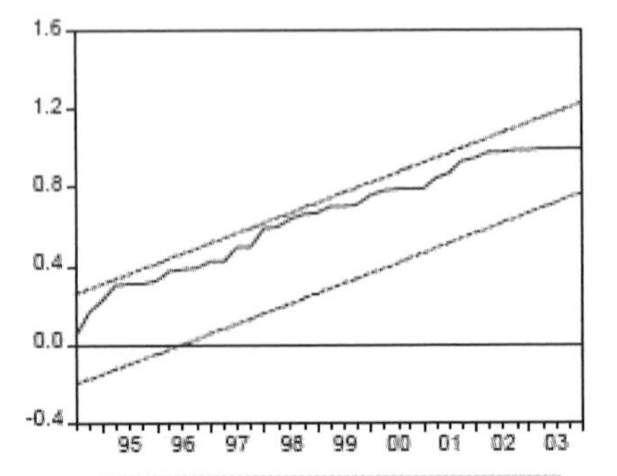

Table 2a. Poland, 1994 onwards.

AD [sample: 94Q1–04Q1]

$\hat{y}_t = 0.020 + 0.467 y_{t-1} - 0.0027 r_{t-1}$

[N: 0.00] [AC: 0.047] [H: 0.49] [Chow: 0.046] [W: 0.57]

PC [sample: 94Q1–04Q1]

$\hat{\Delta\pi} = -0.64\Delta\pi_{t-1} - 0.67\Delta\pi_{t-2} - 0.51\Delta\pi_{t-3} + 21.46(q_{t-1} - q_{t-5})$

[N: 0.00] [AC: 0.056] [H: 0.29] [Chow: 0.36] [W: 0.06] [Chow: 2000$Q$2: 0.65]

CUSUMsq, AD

CUSUMsq, PC

Table 2b. Poland, restricted sample.

AD [sample: 95$Q$1–04$Q$1]

$\hat{y}_t = 0.021 + 0.454 y_{t-1} - 0.0029 r_{t-1}$

[N: 0.00] [AC: 0.07] [H: 0.54] [Chow: 0.54] [W: 0.64]

PC [sample: 94$Q$1–04$Q$1]

$\hat{\Delta\pi}_t = -0.57\Delta\pi_{t-1} - 0.76\Delta\pi_{t-2} - 0.47\Delta\pi_{t-3} + 16.09(q_{t-1} - q_{t-5})$

[N: 0.57] [AC: 0.25] [H: 0.79] [W: 0.39]

CUSUMsq, AD

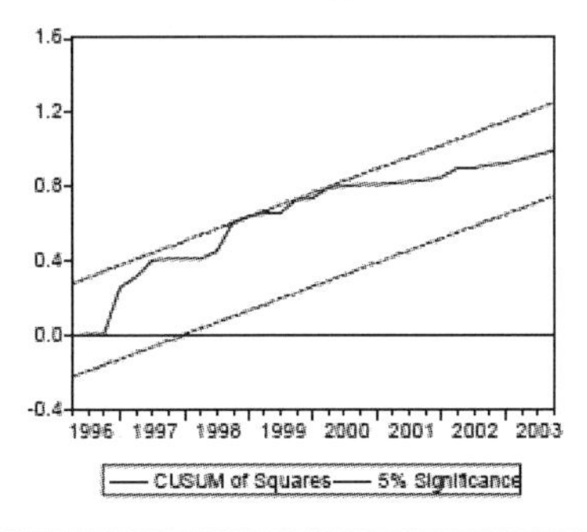

Table 3. Czech Republic, restricted sample.

AD [sample: $98Q1$–$04Q1$]

$$\hat{y}_t = \underset{(0.006)}{0.010} + \underset{(0.116)}{0.620} y_{t-1} - \underset{(0.001)}{0.03} r_{t-1} + \underset{(0.050)}{0.083} (q_{t-1} - q_{t-5})$$

[N: 0.79] [AC: 0.07] [H: 0.54] [W: 0.15]

PC [sample: $98Q1$–$04Q1$]

$$\hat{\Delta\pi}_t = -0.76\Delta\pi_{t-1} - 0.21\Delta\pi_{t-2} - 0.31\Delta\pi_{t-3} + 32.55(q_{t-1} - q_{t-5})$$

[N: 0.71] [AC: 0.34] [H: 0.33] [W: 0.26]

Table 4. Slovenia.

AD [sample: $94Q1$–$03Q3$]

$$\hat{y}_t = \underset{(0.167)}{0.226} y_{t-1} + \underset{(0.234)}{0.343} w_{t-1}$$

[N: 0.30] [AC: 0.37] [H: 0.86] [Chow 0.31] [W: 0.31]

PC [sample: $94Q1$–$03Q4$]

$$\hat{\Delta\pi}_t = -0.60\Delta\pi_{t-1} - 0.66\Delta\pi_{t-2} - 0.45\Delta\pi_{t-3} + 52.64(q_{t-1} - q_{t-5})$$

[N: 0.44] [AC: 0.35] [H: 0.65] [Chow: 0.83] [W: 0.15] [Break $2001Q1$:0.69]

CUSUMsq, AD

CUSUMsq, PC

Datastream codes.

	Germany	**Poland**	**Czech Republic**	**Slovenia**
Exchange rate (US$ per €)	USXRUSD			
Exchange rate (vs. US$)	BDESXUSD.	POI..AF.	CZXRUSD.	SJXRUSD.
CPI	BDCONPRCF	POCONPRCF	CZCONPRCF	SJCONPRCF
Industrial production	BDINPRODG	POIPTOT5H	CZIPTOT.H	SJIPTO01H
3-month-rate	BDINTER3	PO160C..	CZ160C..	SJI60B..

Diagnostic tests of some specifications that we dismissed:

Table 5b. Germany, PC equation, whole dataset.

PC [sample: $91Q3$–$04Q1$]

$$\hat{\Delta\pi}_t = \underset{(0.48)}{0.47} - \underset{(0.18)}{0.29}\pi_{t-1} - \underset{(0.18)}{0.66}\Delta\pi_{t-1} - \underset{(0.17)}{0.53}\Delta\pi_{t-2} - \underset{(0.18)}{0.50}\Delta\pi_{t-3}. + \underset{(13.41)}{7.22}y_{t-1}$$

[N:0.00] [AC: 0.20] [H: 0.06] [Chow: 0.62]

CUSUMsq, PC

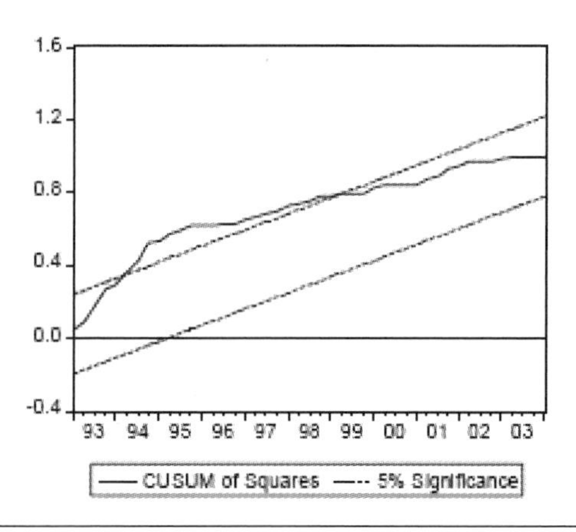

Table 5c. Poland, whole dataset.

AD [sample: 93Q2–04Q1]

$$\hat{y}_t = \underset{(0.009)}{0.02} + \underset{(0.166)}{0.509y_{t-1}} - \underset{(0.001)}{0.0025r_{t-1}} - \underset{(0.203)}{0.130w_{t-1}} - \underset{(0.045)}{0.054(q_{t-1} + q_{t-5})}$$

[N: 0.00] [AC: 0.20] [H: 0.53] [Chow: 0.06]

PC [sample: 93Q2–04Q1]

$$\hat{\Delta\pi}_t = -\underset{(1.25)}{0.42} - \underset{(0.08)}{0.10\pi_{t-1}} - \underset{(0.12)}{0.77\Delta\pi_{t-1}} - \underset{(0.11)}{0.78\Delta\pi_{t-2}}$$

$$- \underset{(0.10)}{0.67\Delta\pi_{t-3}}. + \underset{(27.42)}{22.55y_{t-1}} + \underset{(9.29)}{18.08(q_{t-1} - q_{t-5})}$$

[N: 0.00] [AC: 0.01] [H: 0.07] [Chow: 0.81]

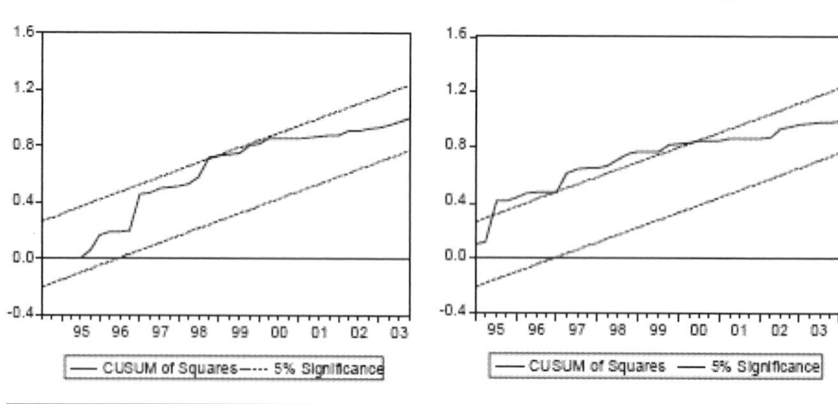

Table 5d. Czech Republic, whole dataset.

AD [sample: 93$Q$4–04$Q$1]

$$\hat{y}_t = \underset{(0.005)}{0.007} + \underset{(0.140)}{0.617} y_{t-1} - \underset{(0.001)}{0.0001} r_{t-1} - \underset{(0.229)}{0.186} w_{t-1} + \underset{(0.063)}{0.047} (q_{t-1} + q_{t-5})$$

[N: 0.44] [AC: 0.94] [H: 0.06] [Chow: 0.01]

PC [sample: 94$Q$1–04$Q$1]

$$\hat{\Delta\pi}_t = \underset{(1.17)}{2.45} - \underset{(0.19)}{0.24} \pi_{t-1} - \underset{(0.21)}{0.68} \Delta\pi_{t-1} - \underset{(0.19)}{0.28} \Delta\pi_{t-2}$$

$$- \underset{(0.10)}{0.36} \Delta\pi_{t-3} + \underset{(24.4)}{22.10} y_{t-1} + 33.22 (q_{t-1} - q_{t-5})$$

[N: 0.00] [AC: 0.25] [H: 0.086 [Chow: 0.01] [W:0.19]

CUSUMsq, AD

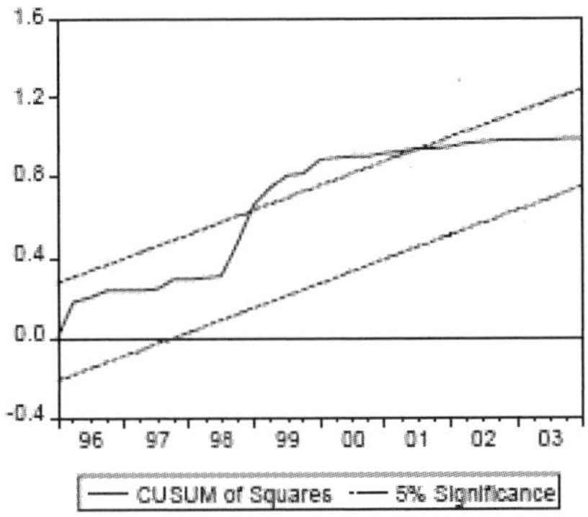

CUSUMsq, PC

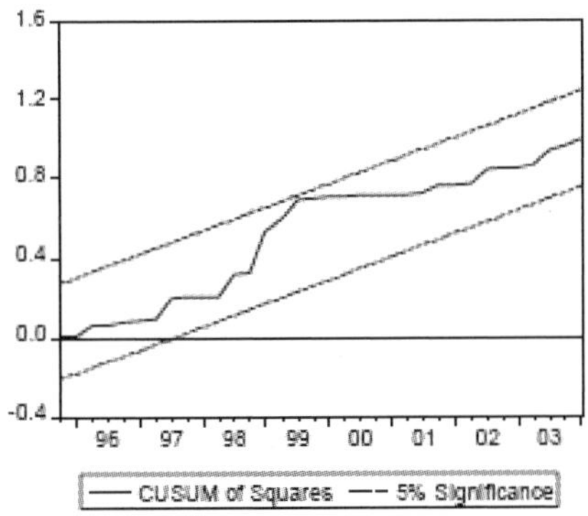

Table 5e. Slovenia, PC equation, whole dataset.

PC [sample: 93$Q$2–03$Q$4]

$$\hat{\Delta}\pi_t = \underset{(1.41)}{4.71} - \underset{(0.19)}{0.50}\pi_{t-1} - \underset{(0.11)}{0.03}\Delta\pi_{t-1} - \underset{(0.11)}{0.40}\Delta\pi_{t-2} - \underset{(0.07)}{0.10}\Delta\pi_{t-3}. + \underset{(26.70)}{2.45}y_{t-1} + \underset{(17.73)}{48.28}(q_{t-1} - q_{t-5})$$

[N: 0.84] [AC: 0.03] [H: 0.02] [Chow: 0.71]

CUSUMsq, PC

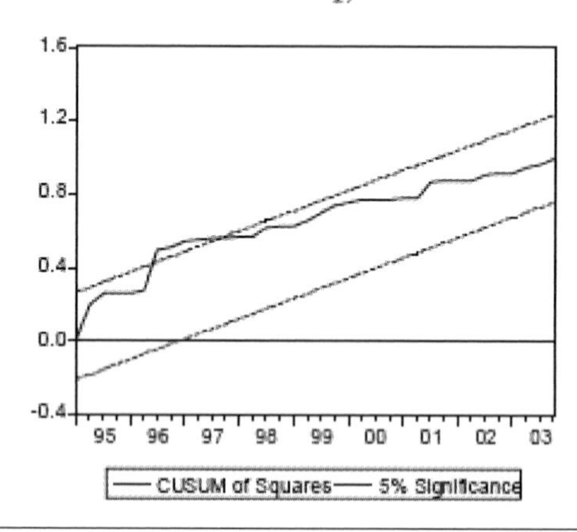

Topic 2

Credit and Income: Co-Integration Dynamics of the US Economy

Athanasios Athanasenas

Institute of Technology and Education, Greece

1. Introduction

In this study, I seek to identify the existence of a significant statistical relationship between credit (as commercial bank lending) and income (GDP), for the US post-war economy, in terms of a contemporary co-integration methodology.

The main empirical finding regarding the hypothesis that causality, in the long run, is directed from credit to income, agrees with the "*credit-view*". This can be verified by the the post-war US data. In this chapter, I have identified a dynamic causality effect in the short run, from income changes to credit ones. To obtain the results cited here, I have applied a co-integration analysis. In addition to considering the formulated VAR, I place particular emphasis upon the stability analysis of the estimated equivalent first-order dynamic system. Finally, dynamic forecasts from the corresponding error correction variance (ECVAR) are obtained, in order to verify the validity of the co-integration vector used.

This study have four sections. Section 2 reviews, in a brief note, the literature on credit and income growth of the post-war US economy. Section 3 deals with methodological issues concerning the contemporary co-integration analysis and the data used. Section 4 presents the econometric analysis and the empirical results; while Section 5 concludes the credit-income causality issue, placing emphasis on the stability analysis of the obtained first-order dynamic system and the forecasts from the corresponding ECVAR.

2. The Credit Issue Researched: A Brief Note on the US Economy

There is an evidence that money tends to "lead" income in a historical sense (see Friedman and Schwartz (1963) and Friedman (1961; 1964). Where the mainstream monetarists through the "quantity theory" explain the empirical observations as the causal relationship running from money to income. Since the seminal works of Sims, empirical validation of the relationship between money and income fluctuations has been established (Sims, 1972; 1980a). Since the publication of the works of Goldsmith (1969), Mckinnon (1973), and Shaw (1973), the debate has been going on about the role of financial intermediaries and bank credit in promoting the long-run growth.

On the basis of historical data, Friedman and Schwartz (1963) argue that major depressions of the US economy have been caused by autonomous movements in money stock.[1] Sims (1980a) concludes that on one hand, money stock emerges as causally prior accounting for a substantial fraction of income variance during the inter-war and post-war periods of the business cycles. On the other hand, old skepticism remains on the direction of causality between money and income.

It is well known that traditional monetarists are consistent with the old-classic view that only "money matters", so that even when tightening of bank credits does occur during recessions, they see these as part of the endogenous financial system, rather than exogenous events, which induce recessions. As Brunner and Meltzer (1988) argue, banking crises are endogenous financial forces, which directly affect business cycles conditional upon the monetary propagation mechanism.[2]

In contrast, the "new-credit" viewers stress further the importance of credit restrictions, while accepting the fundamental inefficiencies of the monetary policy.[3] Bernanke and Blinder (1988), in their seminal article, conclude that credit demand is becoming relatively more stable than money

---

[1] *Ibid.*

[2] Note that monetarists seem to stress mainly, the complementarity of credit and money channels of transmission, the short-run stability of the money multiplier, and the long-run neutrality of money.

[3] The "new-credit" view combines the "old-credit" view of the money to spend perspective, with the money to hold perspective of the money view. The "old-credit" view focused on decisions to create and spend money, on the expansion effects of bank lending, and emphasized the issue of the non-neutrality of credit money in the long-run (see also, Trautwein, 2000).

demand since 1979 and throughout the 1980s, where, money demand shocks became much more important relative to credit demand shocks in the 1980s, supported by their estimation of greater variance of money.

Evidently, one of the most challenging issues of the applied monetary theory is the existence and importance of the credit channel in the monetary transmission mechanism. Bernanke (1988) emphasizes that according to the proponents of the credit view, bank credit (loans) deserve special treatment due to the qualitative difference between a borrower going to the bank for a loan and a borrower raising funds by going to the financial markets and issuing stock or floating bonds.[4]

Kashyap and Stein (1994) leave no doubt while supporting Bernanke's (1983), and Bernanke's and James's (1991) examination of the Great Depression in the United States, that the conventional explanation for the depth and persistence of the Depression is one of the strongest pieces of evidence supporting the view that shifts in loan supply can be quite important. The supporting evidence stands strong, considering also that the decline of intensity as seen in recessions are partially due to a cut-off in bank lending (Kashyap *et al.*, 1994).[5]

King and Levine (1993) stress that financial indicators, such as the importance of the banks relative to the central bank, the percentage of credit allocated to private firms and the ratio of credit issued to private firms to GDP, are strongly related with growth. For the relevant role of the monetary process and the financial sector, see, in particular, (Angeloni *et al.*, 2003; Bernanke and Blinder, 1992; Murdock and Stiglitz, 1993; Stock and Watson, 1989; Swanson, 1998).

More recently, Kashyap and Stein (2000) proved that the impact of monetary policy on lending behavior is stronger for the banks with less liquid balance sheets, where liquidity is measured by the ratio of securities

---

[4]Stiglitz (1992) stresses the fact that the distinctive nature of the mechanisms by which credit is allocated plays a central role in the allocation process. He identifies several crucial reasons that banks are likely to reduce lending activity (p. 293), as the economy goes into recession, where banks' unwillingness and the inability to make loans obviates the effect of monetary policy. Bernanke and Blinder (1992) find that the transmission of monetary policy works through bank loans, as well as through bank deposits. Tight monetary policy reduces deposits and over time banks respond to tightened money by terminating old loans and refusing to make new ones. Reduction of bank loans for credit can depress the economy.

[5]Both Bernanke and Blinder (1992) and Gertler and Gilchrist (1993) using innovations in the federal funds rate as an indicator of monetary policy disturbances, find that M1 or M2 declines immediately, while bank loans are slower to fall, moving contemporaneously with output.

to assets. Their principal conclusions can be narrowed down to: (a) the existence of the credit-lending channel of monetary transmission, at least for the United States is undeniable and (b) the precise quantification and accurate measurement of the aggregate loan-supply consequences of monetary policy remain very difficult in econometrics due to the large estimation biases. See also the works of Mishkin (1995), Bernanke and Gertler (1995), and Meltzer (1995), for the bank-lending channel. Lown and Morgan (2006) stress further the substantial issue of the role of fluctuations in commercial credit standards and credit cycles being correlated with real output.

All previous works provide a clear evidence that is consistent with the existence of a *credit-lending channel* of monetary transmission. The very heart of the *credit-lending view* is the proposition that the Central Banks, in general, and the Federal Reserve in the United States, in particular, can shift banks' loan supply schedules, by conducting open-market operations.[6]

Consequently, given the afore-mentioned empirical research on the credit-lending channel and the established relationship between financial intermediation and economic growth in general, our main focus here is to *analyze in detail* the causal relationship between finance and growth by focusing on bank credit and income GNP, in the post-war US economy. *Analyzing in detail* here means placing particular emphasis: (a) on a contemporary co-integration approach, and (b) on the stability analysis of the observed first-order dynamic system and the dynamic forecasts from the corresponding ECVAR.

3. Methodological Issues and Data

In the empirical analysis, for the case of the US post-war period 1957–2007, we test for the existence of causality between the total loans and leases at commercial banks, denoted by *real credit*, representing the development of the banking sector, and the money income, GNP, denoted by *real income*, representing national economic performance. Both the variables are quarterly seasonally adjusted from 1957:3 to 2007:3, i.e., 201 observations, and are expressed in real terms, indexed by 1982–1984 as base 100, in billions of US$. Considering the natural logs, I define two new variables (W_i, Y_i) such that $Y_i = \ln(\textit{real income}_i)$ and $W_i = \ln(\textit{real credit}_i)$, for $i = 1, 2, \ldots, 201$.

---

[6]It is noted that the credit-lending channel also requires an imperfect price adjustment and that some borrowers cannot find perfect substitutes for bank loans.

My data is obtained on line from the Federal Reserve Bank of Saint Louis, Missouri, through their open source database.[7]

Researchers usually employ some measure of the stock of money over GNP, as a proxy for the size of the "financial depth" (see for instance, Goldsmith, 1969; Mckinnon, 1973). This type of financial indicator does not inform us as to whether the liabilities are those of the central banks, commercial banks, or other financial intermediaries. In this way, the credit-channel issue cannot be analyzed, and the significant role of credit cannot be isolated. Moreover, this type of financial indicator is mainly applicable to the developing countries, since the financing of the economy is to a great extent carried out by the public sector and controlled mainly by its central bank.

An advanced economy such as the United States, with a fully liberalized banking and financial system, provides an excellent case for testing the modern credit view such that the commercial bank credit itself has to be modeled as a fundamental, financial, and an independent monetary variable that affects money income, and hence the economic performance, in the long run.

Based on the cited references, we would accept the value of commercial bank credit to the private sector, as a measure of the ability of the banking system to provide finance-led income growth. The extent to which money income growth is associated with the credit provision from the financial sector is examined through the use of contemporary co-integration methods and system stability analysis.

Initially, the order of integration of the two variables is investigated using standard unit root tests (Dickey and Fuller, 1979; Harris, 1995). The next step is the computation of co-integration vectors by applying the maximum likelihood (ML) approach considering the error-correction representation of the VAR model (Harris, 1995; Johansen and Juselius, 1990). We transformed the initial VAR to an equivalent first-order dynamic system to perform the proper stability analysis, which revealed that the system under consideration is stable. It should be re-called that stability is of vital importance for obtaining consistent forecasts, and this issue is stressed appropriately in Sections 4 (4.2 and 4.6) and 5.

I proceed with the utilization of the vector error-correction modeling following Engle and Granger (1987), and testing for the short-run dynamic relationship on causality effects between money and income.

---

[7]And, we are deeply thankful for this.

I end up by applying a dynamic forecasting technique with the estimated ECVAR. The very low value of Theil's inequality coefficient U, provides a concrete evidence that I have formulated the suitable ECVAR obtaining, thus, robust results. Our particular emphasis on the system stability is evident and it is clarified in detail in the following section.

4. Co-Integration

Considering the series $\{Y_i\}$, $\{W_i\}$ and applying the Dickey–Fuller (DA/ADF) tests (Dickey and Fuller, 1979; 1981; Harris, 1995, pp. 28–47), I found that both series are not stationary, but they are integrated of order 1, that is $I(1)$. This implies that by differentiating the initial series, we get stationary ones. In other words, the series $\Delta Y_i = Y_i - Y_{i-1}$ and $\Delta W_i = W_i - W_{i-1}$ are stationary, that is $I(0)$. This can be easily verified from Fig. 1, where the non-stationary series $\{Y_i\}$, $\{W_i\}$ are also presented.

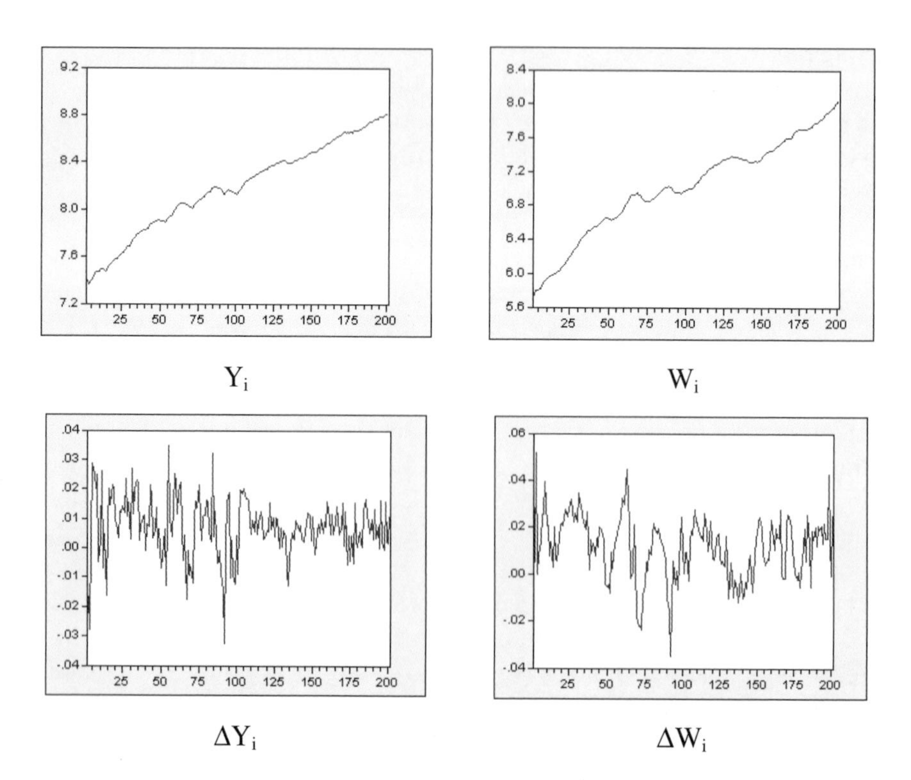

Figure 1. The $I(1)$ series $\{Y_i\}$, $\{W_i\}$ and the stationary ones $\{\Delta Y_i\}$, $\{\Delta W_i\}$.

Since the variables Y_i, W_i are used in this study, it should be re-called that the parameters of a long-run relationship represent constant elasticities.

It should be re-called that if two series are not stationary but they have common trends so that there exist a linear combination of these series that is stationary, which means that they are integrated in a similar way and for this reason, they are called co-integrated, in the sense that one or the other of these variables will tend to adjust so as to restore a long-run equilibrium. Also, if there is a linear trend in the data, as in the case of Y_i and W_i as shown in Fig. 1, then we specify a model including a time-trend variable, allowing thus the non-stationary relationships to drift.

4.1. *The VAR Model and The Equivalent ECVAR*

We start this section by formulating and estimating an unrestricted VAR of the form:

$$\mathbf{x}_i = \boldsymbol{\delta} + \boldsymbol{\mu} t_i + \sum_{j=1}^{p} \mathbf{A}_j \mathbf{x}_{i-j} + \mathbf{w}_i \tag{1}$$

where $\mathbf{x} \in E^n$ (here E denotes the Euclidean space), $\boldsymbol{\delta}$ is the vector of constant terms, t_i is a time-trend variable, $\boldsymbol{\mu}$ is the vector of corresponding coefficients, and $\mathbf{w}$ is the noise vector. The matrices of coefficients $\mathbf{A}_j$, are defined on $E^n \times E^n$. It is noted that variable t_i is included for the reason mentioned above. It can be easily shown (Johansen, 1995), that (1) may be expressed in the form of an equivalent ECVAR, that is:

$$\Delta\mathbf{x}_i = \boldsymbol{\delta} + \boldsymbol{\mu} t_i + \sum_{j=1}^{p-1} \mathbf{Q}_j \Delta\mathbf{x}_{i-j} + \Pi\mathbf{x}_{i-1} + \mathbf{w}_i \tag{2}$$

where $\mathbf{Q}_j = (\sum_{k=1}^{j} \mathbf{A}_k - \mathbf{I})$.

After estimation, we may compute matrix $\mathbf{\Pi}$ from:

$$\mathbf{\Pi} = \left(\sum_{j=1}^{p} \mathbf{A}_j \right) - \mathbf{I} \tag{2a}$$

or, directly from the corresponding ECVAR as given in Eq. (2). It is re-called that this type of VAR models like the one presented above, is recommended by Sims (1980a; b), as a way to estimate the dynamic relationships among jointly endogenous variables. It should be noted that for the case under consideration, vector $\mathbf{x}$ in Eqs. (1) and (2) is 2-dimensional, i.e.,

$$\mathbf{x}_i = \begin{bmatrix} Y_i \\ W_i \end{bmatrix} \text{ and it is } \mathbf{x} \sim I(1).$$

It is known that the matrix $\mathbf{\Pi}$ can be decomposed such that $\mathbf{\Pi} = \mathbf{AC}$, where the elements of $\mathbf{A}$ are the coefficients of adjustment and the rows of matrix $\mathbf{C}$ are the (possible) co-integrating vectors. In some books, $\mathbf{A}$ is denoted by $\boldsymbol{\alpha}$ and $\mathbf{B}$ by $\boldsymbol{\beta}'$, although capital letters are used for matrices. Besides, $\boldsymbol{\beta}$ denotes the coefficient vector in a general linear econometric model. To avoid any confusion, we adopted the notation used here, as well as in Mouza and Paschaloudis (2007).

Usually, matrices $\mathbf{A}$ and $\mathbf{C}$ are obtained by applying the ML method (Harris, 1995; Johansen and Juselius, 1990, p. 78). According to this procedure, a constant and/or trend can be included in the co-integrating vectors in the following way.

Considering Eq. (2), we can augment matrix $\mathbf{\Pi}$ to accommodate, as an additional column, the vectors $\boldsymbol{\delta}$ and $\boldsymbol{\mu}$ in the following way.

$$\tilde{\Pi} = [\mathbf{\Pi} \vdots \boldsymbol{\mu}\boldsymbol{\delta}]. \tag{3}$$

In this case, $\tilde{\Pi}$ is defined on $E^n \times E^m$, with $m > n$. For conformability, the vector $\mathbf{x}_{i-1}$ in Eq. (2) should be augmented accordingly by using the linear advance operator L (such that $L^k y_i = y_{i+k}$), i.e.,

$$\tilde{\mathbf{x}}_{i-1} = \begin{bmatrix} \mathbf{x}_{i-1} \\ Lt_{i-1} \\ 1 \end{bmatrix}. \tag{3a}$$

Hence, Eq. (2) takes the form:

$$\Delta\mathbf{x}_i = \sum_{j=1}^{p-1} \mathbf{Q}_j \Delta\mathbf{x}_{i-j} + \tilde{\Pi}\tilde{\mathbf{x}}_{i-1} + \mathbf{w}_i. \tag{4}$$

Note that no constants (vector $\boldsymbol{\delta}$), as well as coefficients of the time trend (vector $\boldsymbol{\mu}$) are explicitly presented in the ECVAR in Eq. (4), which specifies precisely a relevant set of n ECM.

It is assumed at this point that matrices $\mathbf{C}$ and $\mathbf{A}$ refer to matrix $\tilde{\Pi}$ as seen in Eq. (4). Let us consider now the kth row of matrix $\mathbf{C}$, that is $\mathbf{c}'_{k.}$.[8] If this row is assumed to be a co-integration vector, then the (disequilibrium)

[8] Note that dot is necessary to distinguish the kth row of $\mathbf{C}$, i.e., $\mathbf{c}'_{k.}$, from the transposed of the kth column of this matrix that is $\mathbf{c}'_{k}$.

errors computed from $u_i = \mathbf{c}'_{k.}\tilde{\mathbf{x}}_i$ must be a stationary series. Considering now the kth column of matrix $\mathbf{A}$, that is $\mathbf{a}_k$, and given that $u_{i-1} = \mathbf{c}'_{k.}\mathbf{x}_{i-1}$, then the ECVAR in Eq. (4) can be written as:

$$\Delta \mathbf{x}_i = \sum_{j=1}^{p-1} \mathbf{Q}_j \Delta \mathbf{x}_{i-j} + \mathbf{a}_k u_{i-1} + \mathbf{w}_i. \tag{5}$$

This is the conventional form of an ECVAR, when matrix $\tilde{\mathbf{\Pi}}$ instead of $\mathbf{\Pi}$ is considered. In any case, the maximum lag in Eq. (5) is $p - 1$. It should be noted that this specification of an ECM is fully justified from the theoretical point of view. However, if we want to include an intercept in Eq. (5), i.e., in the short-run model, it may be assumed that the intercept in the co-integration vector is cancelled by the intercept in the short-run model, leaving only an intercept in Eq. (5). Thus, when a vector of constants is to be included in Eq. (4), which, in this case, takes the form:

$$\Delta \mathbf{x}_i = \boldsymbol{\delta} + \sum_{j=1}^{p-1} \mathbf{Q}_j \Delta \mathbf{x}_{i-j} + \tilde{\mathbf{\Pi}} \tilde{\mathbf{x}}_{i-1} + \mathbf{w}_i \tag{6}$$

then Eq. (5), matrix $\tilde{\mathbf{\Pi}}$ and the augmented vector $\tilde{\mathbf{x}}$ will become:

$$\Delta \mathbf{x}_i = \boldsymbol{\delta} + \sum_{j=1}^{p-1} \mathbf{Q}_j \Delta \mathbf{x}_{i-j} + \mathbf{a}_k u_{i-1} + \mathbf{w}_i \tag{6a}$$

$$\tilde{\mathbf{\Pi}} = [\mathbf{\Pi} \vdots \boldsymbol{\mu}] \tag{6b}$$

$$\tilde{\mathbf{x}}_{i-1} = \begin{bmatrix} \mathbf{x}_{i-1} \\ Lt_{i-1} \end{bmatrix}. \tag{6c}$$

According to Eq. (6a), the co-integrating vector $\mathbf{c}'_{k.}$, used to compute $u_i = \mathbf{c}'_{k.}\tilde{\mathbf{x}}_i$, includes only the coefficient of the time-trend variable. The inclusion of a trend in the model can be justified, if we examine the estimation results of a long-run relationship considering Y and W. In such a case, we immediately realize that a trend should be present.

After adopting the VAR as seen in Eq. (1), the next step is to determine the value of p from Table 1 (Holden and Perman, p. 108).

We see from Table 1 that for $\alpha \leq 0.05$, $p = 4$, which means that we have to estimate a VAR(4), with only one deterministic terms (i.e., trend). After estimation, we found that matrix $\tilde{\mathbf{\Pi}}$ (hat is omitted for simplicity),

Table 1. Determination of the value of p.

Lag length p	LR	Value of p (probability)
1	—	—
2	132.91	0.0000
3	14.988	0.0047
4	14.074	0.0071
5	2.7328	0.6035
6	5.1465	0.2726
7	2.3439	0.6728
8	5.2137	0.2661

specified in Eq. (6b) has the following form:

$$\tilde{\Pi} = \begin{bmatrix} \overset{Y_i}{-0.092971} & \overset{W_i}{0.028736} & \overset{t_i}{0.000320} \\ 0.016209 & -0.024594 & 0.000124 \end{bmatrix}. \tag{7}$$

4.2. *Dynamic System Stability*

The VAR(4) can be transformed to an equivalent first-order dynamic system, in the following way.

$$\begin{bmatrix} \mathbf{x}_i \\ L\mathbf{x}_i \\ L^2\mathbf{x}_i \\ L^3\mathbf{x}_i \end{bmatrix} = \begin{bmatrix} \mathbf{A}_1 & \mathbf{A}_2 & \mathbf{A}_3 & \mathbf{A}_4 \\ \mathbf{I}_2 & \mathbf{0} & \mathbf{0} & \mathbf{0} \\ \mathbf{0} & \mathbf{I}_2 & \mathbf{0} & \mathbf{0} \\ \mathbf{0} & \mathbf{0} & \mathbf{I}_2 & \mathbf{0} \end{bmatrix} \begin{bmatrix} \mathbf{x}_{i-1} \\ L\mathbf{x}_{i-1} \\ L^2\mathbf{x}_{i-1} \\ L^3\mathbf{x}_{i-1} \end{bmatrix} + \begin{bmatrix} \boldsymbol{\delta} \\ \mathbf{0} \\ \mathbf{0} \\ \mathbf{0} \end{bmatrix} + \begin{bmatrix} \boldsymbol{\mu} \\ \mathbf{0} \\ \mathbf{0} \\ \mathbf{0} \end{bmatrix} t_i + \begin{bmatrix} \mathbf{w}_i \\ \mathbf{0} \\ \mathbf{0} \\ \mathbf{0} \end{bmatrix}. \tag{8}$$

L in this place, denotes the linear lag operator, such that $L^k z_i = z_{i-k}$. Equation (8) can be written in a compact form as:

$$\mathbf{y}_i = \tilde{\mathbf{A}}\mathbf{y}_{i-1} + \mathbf{d} + \mathbf{z}t_i + \mathbf{v}_i \tag{8a}$$

where

$$\mathbf{y}_i' = [\mathbf{x}_i \quad L\mathbf{x}_i \quad L^2\mathbf{x}_i \quad L^3\mathbf{x}_i], \quad \mathbf{d}' = [\boldsymbol{\delta} \quad \mathbf{0} \quad \mathbf{0} \quad \mathbf{0}],$$
$$\mathbf{z}' = [\boldsymbol{\mu} \quad \mathbf{0} \quad \mathbf{0} \quad \mathbf{0}], \quad \mathbf{v}_i' = [\mathbf{w}_i \quad \mathbf{0} \quad \mathbf{0} \quad \mathbf{0}]$$

and the dimension of matrix $\tilde{\mathbf{A}}$ is (8×8).

Table 2. Eigen values.

No.	Real part	Coefficient of imaginary part	Length
1	0.94	0	0.94
2	0.8567	0.0579	0.8586
3	0.8567	−0.0579	0.8586
4	0.3807	0.0	0.3807
5	0.0367	0.3655	0.3674
6	0.0367	−0.3677	0.3674
7	−0.3581	0.0	0.3581
8	−0.1142	0.0	0.1142

To decide whether the dynamic system described by Eq. (8a) is stable, we compute the eigenvalues of matrix $\tilde{\mathbf{A}}$, presented in Table 2.

Since the greater length (0.94) is less then 1, the dynamic system as given in Eq. (8a) is stable and can be used for forecasting purposes. A side verification of this stability test, is that once the system is stable, then the elements of the state vector $\mathbf{x}$ in the initial VAR(4), which in this case are the series $\{Y_i\}$, $\{W_i\}$, are either difference stationary series (DSS), or in some cases, stationary series. In fact, we show that these series are DSS and in particular, they are $I(1)$.

4.3. *The Estimated Co-Integrating Vectors*

We applied the ML method, to compute matrices **C** and **A**, which have the following form:

$$\begin{array}{ccc} & Y \quad\quad W \quad\quad\quad\quad t & \\ \mathbf{C} = & \begin{bmatrix} 1 & -0.289613 & -0.003621 \\ 1 & -0.653108 & -0.000292 \end{bmatrix}, & \\ \mathbf{A} = & \begin{bmatrix} -0.087991 & -0.004980 \\ -0.038535 & 0.054745 \end{bmatrix} & \end{array} \quad (9)$$

satisfying $\mathbf{AC} = \tilde{\Pi}$. We can easily verify that only the first row of matrix **C** is a promising candidate, verified from the computed value of the t-statistic (−0.23), which corresponds to the (1, 2) element of matrix **A**, that corresponds to the second row of matrix **C**. Thus, we have the co-integrating vector:

$$[1 \quad -0.289613 \quad -0.003621] \tag{10}$$

producing the errors u_i, which are the disequilibrium errors obtained from:

$$u_i = Y_i - 0.289613W_i - 0.003621t_i. \tag{10a}$$

It should be noted that the vector specified in Eq. (10) is a co-integration vector, iff the series $\{u_i\}$ computed from Eq. (10a) is stationary. To find out that this series is stationary, we run the following regression with a constant term, since $\sum u_i \neq 0$.

$$\Delta u_i = a + b_1 u_{i-1} + \sum_{j=1}^{q} b_{j+1}\Delta u_{i-j} + \varepsilon_i. \tag{11}$$

Note that the value of q is set such that the noises ε_i to be white. With $q = 3$, the estimation results are as follows:

$$\Delta u_i = \underset{(0.1174)}{0.4275} - \underset{(0.02029)}{0.07384}\, u_{i-1} + \sum_{j=1}^{3} \hat{b}_{j+1}\Delta u_{i-j} + \hat{\varepsilon}_i \tag{11a}$$
$$t = -3.64.$$

It should be noted that in Eq. (11a), there is no problem regarding autocorrelation and heteroscedasticity. We have to compute the t_u-statistic from MacKinnon (1991, pp. 267–276) critical values (see also Granger and Newbold, 1974; Hamilton, 1994; Harris, 1995, Table A6, p. 158; Harris, 1995, pp. 54–55). This statistic ($t_u = \Phi_\infty + \Phi_1/T + \Phi_2/T^2$, where T denotes the sample size) is evaluated from the relevant table, taking into account Eqs. (10a) and (11a). Note that in the latter equation, there is only one deterministic term (constant). The value of the t_u-statistic for this case is -3.3676 ($\alpha = 0.05$). Hence, since $-3.64 < -3.3676$, we reject the null that the series $\{u_i\}$ is not stationary. Recall that the null $[u_i \sim I(1)]$ is rejected in favor of H_1 $[u_i \sim I(0)]$, if $t < t_u$. Hence, the first row of matrix **C** is a co-integration vector, which yields the following long-run relationship:

$$Y_i = 0.289613W_i + 0.003621t_i + u_i$$

which implies that a 10% increase of W (total credit) results to an increase of Y (money income) by about 2.9%.[9]

---

[9]Another point of interest is that in Eq. (10a), the errors u_i are computed from:

$$u_i = Y_i - \hat{Y}_i \tag{12}$$

where $\hat{Y}_i = 0.289613W_i + 0.003621t_i$.

4.4. *Essential Remarks*

In seminal works about money and income (as for instance in Tobin, 1970), the analysis is restricted in a pure theoretical consideration. Some authors (see for instance, Friedman and Kuttner, 1992) write many explanations about co-integration, co-integrating vectors, and error-correction models, but I did not trace any attempt to compute either. In a later paper by the same authors (Friedman and Kuttner, 1993), I read (p. 200) the phrase, "*vector autoregression system*," which in fact is understood as a VAR. The point is that no VAR model is specified in this work. Besides, the lag length is determined by simply applying the F-test (p. 191). In this context, the validity of the results is questionable. In other applications (see for instance, Arestis and Demetriades, 1997), where a VAR model has been formulated, nothing is mentioned about the test applied to determine the maximum lag length. Also, in this paper, the authors are restricted to the use of trace statistic ($l_{(\text{trace})}$) to test for co-integration (p. 787), although this is the necessary but not the sufficient condition, as pointed out above. From other research works (see for instance, Arestis *et al.*, 2001), I get the impression that there is not a clear notion regarding stationarity and/or integration, since the authors state [p. 22 immediately after Eq. (1)], that "*D is a set of I(0) deterministic variables such as constant, trend and dummies...*". It is clear that this verification is false. How can a trend, for instance, be a stationary, $I(0)$, variable?[10] The authors declare (p. 24) that "*We then perform co-integration analysis...*". The point is that apart from some theoretical issues, I did not trace such an analysis in the way it is applied here. Besides, nothing is mentioned about any ECVAR, analogous to the one that is formulated and used in this study. Most essentially, I have not traced,

---

In case that the disequilibrium errors are computed from:

$$u_i = \hat{Y}_i - Y_i \tag{13}$$

then the co-integration vector will have the form:

$$[-10.289613 \quad 0.003621]. \tag{14}$$

It is noted also that hat (^) is omitted from the disequilibrium errors u_i, for avoiding any confusion with the OLS disturbances.

It should be emphasized that in such a case, i.e., using the disequilibrium errors computed from Eq. (13), then the sign of the coefficient of adjustment, as will be seen later, will be changed.

[10] It should be re-called at this point, that if t_i is a common trend, then $\Delta t_i = 1 (\forall i)$.

in relevant works, the stability analysis applied for the system presented in Eq. (8).

4.5. *The ECM Formulation*

The lagged values of the disequilibrium errors, that is u_{i-1}, serve as an error correction mechanism in a short-run dynamic relationship, where the additional explanatory variables may appear in lagged first differences. All variables in this equation, also known as ECM, are stationary so that, from the econometric point of view, it is a standard single equation model, where all the classical tests are applicable. It should be noted, that the lag structure and the details of the ECM, should be in line with the formulation as seen in Eq. (6a). Hence, I started from this relation considering the errors u_i and estimated the following model, given that the maximum lag length is $p-1$ i.e., 3.

$$\Delta Y_i = \alpha_0 + \sum_{j=1}^{3} \alpha_j \Delta Y_{i-j} + \sum_{j=1}^{3} \beta_j \Delta W_{i-j} - a_Y u_{i-1} + {}_Y v_i. \tag{15}$$

Note that ${}_Y v_i$ are the model disturbances. If the adjustment coefficient a_Y is significant, then we may conclude that in the long run, W causes Y. If $a_Y = 0$, then no such a causality effect exists. In case that all β_j are significant, then there is a causality effect in the short run, from ΔW to ΔY. If all $\beta_j = 0$, then no such causality effect exists. The estimation results are presented below.

$$\Delta Y_i = 0.5132 + 0.2618\Delta Y_{i-1} + 0.161\Delta Y_{i-2} - 0.00432\Delta Y_{i-3} + 0.0273\Delta W_{i-1}$$

	(0.125)	(0.074)	(0.073)	(0.0738)	(0.075)
p value	0.0001	0.0005	0.0276	0.953	0.719
Hansen	0.0370	0.1380	0.2380	0.234	0.128

$$+0.0487\Delta W_{i-2} - 0.076\Delta W_{i-3} \quad - 0.0879 u_{i-1} + {}_Y\hat{v}_i$$

	(0.075)	(0.065)	(0.022)
p value	0.519	0.242	0.0001
Hansen	0.065	0.098	0.025 (for all coefficients 2.316)

(16)

$$\bar{R}^2 = 0.167,\ s = 0.009,\ DWd = 1.98,\ F_{(7,189)} = 6.65,$$

$$\text{Condition number (CN)} = 635.34.$$

$$(p\ value = 0.0)$$

We observe that the estimated adjustment coefficient (−0.0879) is significant and almost identical to the (1, 1) element of matrix **A**, as seen in Eq. (9), which is computed by the ML method. According to Hansen statistics, all coefficients seem to be stable for $\alpha = 0.05$. The BG test revealed that no autocorrelation problems exist. In models like the one seen in Eq. (16), the term u_{i-1} usually produces the smallest Spearman's correlation coefficient (r_s). In this particular case, we have: $r_s = -0.0979$, $t = -1.37$, $p = 0.172$, which means that no heteroscedasticity problems exist. Finally, the RESET test shows that there is no any specification error.

In Eq. (16), we have an inflated CN due to spurious multicollinearity, which is verified from the revised CN, having the value of 3.42 and as Lazaridis (2007) has shown, this means that, in fact, there is not any serious multicollinearity problem, affecting the reliability of the estimation results. In many applications, particularly when the variables are in logs, then usually the value of this statistic (CN) is extremely high, which in most cases, as stated above, is due to the spurious multicollinearity, which can be revealed by computing the revised CN, which is not widely known and it seems that this is the main reason, for not reporting this statistic in relevant applications.

To test the null, H_0: $\beta_1 = \beta_2 = \beta_3 = 0$, where β_j are the coefficients of ΔW_{i-j} ($j = 1, 2, 3$), we computed the F-statistic that is $F_{(3,189)} = 0.533$ (p- value $= 0.66$). Hence, the null is accepted, which means that there is not any short-run causality effect from ΔW to ΔY, but only in the levels, i.e., in the long run, W causes Y. Observing the value of the adjustment coefficient (−0.0879), I may conclude that it gives a satisfactory percentage, regarding the money income convergence towards a long-run equilibrium.[11]

4.6. *Dynamic Forecasting with an ECVAR*

To complete the ECVAR model, it is necessary to estimate a short-run relation for ΔW. Thus, we consider an equation which is similar to Eq. (15) i.e.,

$$\Delta W_i = b_0 + \sum_{j=1}^{3} a_j \Delta Y_{i-j} + \sum_{j=1}^{3} b_j \Delta W_{i-j} - a_W u_{i-1} + {}_W v_i. \tag{17}$$

[11] It is already mentioned that if the disequilibrium errors are computed from Eq. (13), instead of Eq. (12), then the estimated coefficient of adjustment will have a positive sign.

The estimation results are as follows[12]:

$$\Delta W_i = 0.2241 + 0.0977\Delta Y_{i-1} + 0.2256\Delta Y_{i-2} + 0.199\Delta Y_{i-3} + 0.4864\Delta W_{i-1}$$

	(0.125)	(0.0733)	(0.072)	(0.073)	(0.075)
p value	0.075	0.183	0.002	0.0072	0.000
Hansen	0.075	0.181	0.041	0.7130	0.045

$$+ 0.0123\Delta W_{i-2} + 0.068\Delta W_{i-3} \quad - 0.03852 u_{i-1} + {}_W\hat{v}_i$$

	(0.87)	(0.29)		(0.0216)
p value	0.519	0.242		0.076
Hansen	0.062	0.181	0.075	(for all coefficients 2.268)

$$\bar{R}^2 = 0.541, \quad s = 0.009, \quad DWd = 1.97,$$
$$F_{(7,189)} = 34.0(p\ value = 0.0), \quad \text{CN} = 635.4, \quad \text{Revised CN} = 3.42. \tag{18}$$

Presumably, we are on the right way, since the estimated adjustment coefficient is almost the same as the one computed by the ML method as in the previous case. This is the (2, 1) element of matrix **A** as seen in Eq. (9). Note that this coefficient is significant for $\alpha \geq 0.08$, which means that there is a rather weak causality effect from Y to W in the long run.

To test the null $H_0 : a_1 = a_2 = a_3 = 0$, where the coefficients $a_j(j = 1, 2, 3)$ as seen in Eq. (17), we compute the relevant F-statistic, i.e., $F_{(3,189)} = 8.42$ (p-value $= 0.0$). This implies that indeed there is a causality effect in the short run, from ΔY to ΔW.

Considering the complete ECVAR, i.e., Eqs. (16) and (18), Eq. (10a) is used to compute the series $\{u_i\}$, together with some trivial identities, I form the following system:

$$\Delta Y_i = \hat{\beta}_0 + \sum_{j=1}^{3} \hat{\alpha}_j \Delta Y_{i-j} + \sum_{j=1}^{3} \hat{\beta}_j \Delta W_{i-j} - \hat{a}_Y u_{i-1} + {}_Y\hat{v}_i$$

$$\Delta W_i = \hat{b}_0 + \sum_{j=1}^{3} \hat{a}_j \Delta Y_{i-j} + \sum_{j=1}^{3} \hat{b}_j \Delta W_{i-j} - \hat{a}_W u_{i-1} + {}_W\hat{v}_i$$

[12]According to Hansen statistics, all coefficients seem to be stable for $\alpha = 0.05$. The Spearman's correlation coefficient (r_s) regarding the term u_{i-1}, is: $r_s = -0.075$, $t = -1.051$, $p = 0.294$, which means that we do not have to bother about heteroscedasticity problems. Also the revised CN indicates that no multicollinearity problems exist, in order to affect the reliability of the estimation results.

$$
\begin{aligned}
Y_i &= \Delta Y_i + Y_{i-1} \\
W_i &= \Delta W_i + W_{i-1} \\
u_i &= Y_i - 0.289613 W_i - 0.0036218 t_i.
\end{aligned} \tag{19}
$$

It may be useful to note that there are three identities in the system and the only exogenous variable present in the system is the trend. We may obtain dynamic simulation results from the ECVAR specified in Eq. (19), in order to obtain predictions for Y_i and W_i and at the same time to verify the validity of the co-integration vector used. This can be achieved by first formulating the following deterministic system.

$$
\mathbf{K}_0 \mathbf{y}_i = \mathbf{K}_1 \mathbf{y}_{i-1} + \mathbf{K}_2 \mathbf{y}_{i-2} + \mathbf{K}_3 \mathbf{y}_{i-3} + \tilde{\mathbf{D}} \mathbf{q}_i \tag{20}
$$

where $\mathbf{y}_i = [\Delta \Upsilon_i \quad \Delta W_i \quad u_i \quad Y_i \quad W_i]'$, $\mathbf{q}_i = [t_i 1]'$

$$
\mathbf{K}_0 = \begin{bmatrix} 1 & 0 & 0 & 0 & 0 \\ 0 & 1 & 0 & 0 & 0 \\ 0 & 0 & 1 & -1 & 0.289613 \\ -1 & 0 & 0 & 1 & 0 \\ 0 & -1 & 0 & 0 & 1 \end{bmatrix},
$$

$$
\mathbf{K}_1 = \begin{bmatrix} 0.2618 & 0.0273 & -0.0879 & 0 & 0 \\ 0.0977 & 0.4864 & -0.03852 & 0 & 0 \\ 0 & 0 & 0 & 0 & 0 \\ 0 & 0 & 0 & 1 & 0 \\ 0 & 0 & 0 & 0 & 1 \end{bmatrix},
$$

$$
\mathbf{K}_2 = \begin{bmatrix} 0.161 & 0.0487 & 0 & 0 & 0 \\ 0.2256 & 0.0123 & 0 & 0 & 0 \\ 0 & 0 & 0 & 0 & 0 \\ 0 & 0 & 0 & 0 & 0 \\ 0 & 0 & 0 & 0 & 0 \end{bmatrix},
$$

$$
\mathbf{K}_3 = \begin{bmatrix} -0.00432 & 0.076 & 0 & 0 & 0 \\ 0.199 & -0.068 & 0 & 0 & 0 \\ 0 & 0 & 0 & 0 & 0 \\ 0 & 0 & 0 & 0 & 0 \\ 0 & 0 & 0 & 0 & 0 \end{bmatrix},
$$

and

$$\tilde{\mathbf{D}} = \begin{bmatrix} 0 & 0.5132 \\ 0 & 0.2241 \\ -0.003621 & 0 \\ 0 & 0 \\ 0 & 0 \end{bmatrix}.$$

Pre-multiplying Eq. (20) by $\mathbf{K}_0^{-1}$ we get:

$$\mathbf{y}_i = \mathbf{Q}_1\mathbf{y}_{i-1} + \mathbf{Q}_2\mathbf{y}_{i-2} + \mathbf{Q}_3\mathbf{y}_{i-3} + \mathbf{D}\mathbf{q}_i \tag{21}$$

where $\mathbf{Q}_1 = \mathbf{K}_0^{-1}\mathbf{K}_1$, $\mathbf{Q}_2 = \mathbf{K}_0^{-1}\mathbf{K}_2$, $\mathbf{Q}_3 = \mathbf{K}_0^{-1}\mathbf{K}_3$, and $\mathbf{D} = \mathbf{K}_0^{-1}\tilde{\mathbf{D}}$.

The system given in Eq. (21) can be transformed to an equivalent first-order dynamic system, in a similar way to the one already described earlier [see Eqs. (8) and (8a)]. It is noted that in this case matrix $\tilde{\mathbf{A}}$ is of dimension (15×15). It is important to mention again that we should avoid using a dynamic system, for simulation purposes, that is unstable. Thus, from this system, we can obtain dynamic simulation results for the variables Y_i and W_i. These results are graphically presented (Fig. 2).

The very low value of Theil's inequality coefficient U, for both cases is a pronounced evidence that, apart from computing the indicated co-integration vector, we have also formulated the suitable ECVAR so that the results obtained are undoubtfully robust.

5. Conclusions and Implications

The purpose of this study is to contribute to the empirical investigation of the co-integration dynamics of the credit-income nexus, within the economic growth process of the post-war US economy, over the period from 1957 to 2007.

Utilizing advanced and contemporary co-integration analysis and applying vector ECM estimation, we place special emphasis on forecasting and system stability analysis. I can say that to the best of my knowledge, similar system stability and forecasting analysis, as the one applied here, is very difficult to meet in the relevant literature, after taking into consideration similar research works. See for example, (Arestis and Demetriades, 1997; Arestis *et al.*, 2001; Demetriades and Hussein, 1996; Friedman and Kuttner, 1992; 1993; Levine and Zervos, 1998; Rousseau and Wachtel, 1998; 2000).

My results state clearly that there is no short-run causality effect from credit changes to income changes, but only in the levels, that is in the long

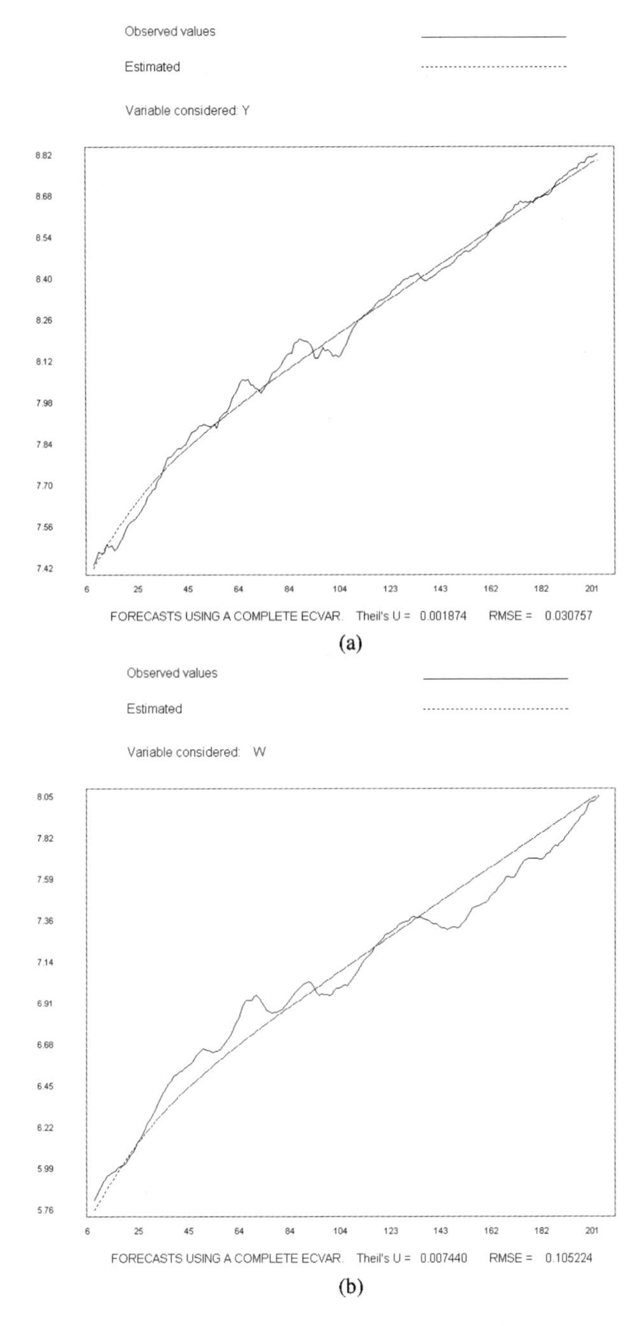

Figure 2. Results of dynamic simulations.

run, credit causes money income. From my co-integration analysis, I found that the adjustment coefficient is significant and its value (−0.0879), gives a satisfactory percentage, regarding the money income convergence towards a long-run equilibrium.

Moreover, we observe a causality effect from income changes to credit changes, in the short run for the post-war US economy. The validity of the credit view theorists seems rather evident, by taking into account the contemporary co-integration and system stability approaches.

Further, in terms of reasonable research implications, following our contemporary co-integration and system stability approaches, we may stress the need for an in-depth investigation of the credit cycle and fluctuations in commercial credit standards.[13]

References

Angeloni, I, KA Kashyap and B Mojon (2003). *Monetary Policy Transmission in the Euro Area*. Part 4, Chap. 24 NY: Cambridge University Press.

Arestis, P and P Demetriades (1997). Financial development and economic growth: assessing the evidence. *The Economic Journal* **107**, 783–799.

Arestis, P, PO Demetriades and KB Luintel (2001). Financial development and economic growth: the results of stock markets. *Journal of Money, Credit and Banking* **33**(1), 16–41.

Bernanke, SB (1983). Non-monetary effects of the financial crisis in the propagation of the great depression. *American Economic Review* **73**(3), 257–276.

Bernanke, SB (1988). Monetary policy transmission: through money or credit? *Business Review*, November/December, 3–11 (Federal Reserve Bank of Philadelphia).

Bernanke, SB and SA Blinder (1988). Credit, money, and aggregate demand. *The American Economic Review, Papers and Proceedings* **78**(2), 435–439.

Bernanke, SB and SA Blinder (1992). The federal funds rate and the channels of monetary transmission. *The American Economic Review* **82**(4), 901–921.

Bernanke, SB and M Gertler (1995). Inside the black box: the credit channel of monetary transmission. *The Journal of Economic Perspectives* **9**(4), 27–48.

Bernanke, SB and H James (1991). The gold standard, deflation, and financial crisis in the great depression: an International comparison. In *Financial Markets and Financial Crises*, RG Hubbard (ed.), Chicago: University of Chicago Press for NBER, 33–68.

Brunner, K and HA Meltzer (1988). Money and credit in the monetary transmission process. *American Economic Review* **78**, 446–451.

Demetriades, PO and K Hussein (1996). Does financial development cause economic growth? *Journal of Development Economics* **51**, 387–411.

[13]See, for example, Lown and Morgan (2006), for parallel research on commercial credit standards, following "traditional" VAR analysis.

Dickey, DA and WA Fuller (1979). Distribution of the estimators for autoregressive time series with a unit root. *Journal of the American Statistical Association* **74**, 427–431.

Dickey, DA and WA Fuller (1981). Likelihood ratio statistics for autoregressive time series with a unit root. *Econometrica* **49**, 1057–1072.

Engle, RF and CWJ Granger (1987). Cointegration and error-correction: representation, estimation, and testing. *Econometrica* **55**, 251–276.

Friedman, M (1961). The lag in the effect of monetary policy. *Journal of Political Economy*, October **69**, 447–466.

Friedman, M (1964). The monetary studies of the national bureau. *National Bureau of Economic Research*, Annual Report. New York: National Bureau of Economic Research, 1–234.

Friedman, B and KN Kuttner (1992). Money, income, prices and interest rates. *The American Economic Review* **82**(3), 472–492.

Friedman, B and KN Kuttner (1993). Another look at the evidence of money-income causality. *Journal of Econometrics* **57**, 189–203.

Friedman, M and A Schwartz (1963). Money and business cycles. *Review of Economics and Statistics*, February, (suppl. 45), 32–64.

Gertler, M and S Gilchrist (1993). The role of credit market imperfections in the monetary transmission mechanism: arguments and evidence. *Scandinavian Journal of Economics* **95**(1), 43–64.

Goldsmith, RW (1969). Financial structure and development. New Haven, CT: Yale University Press.

Granger, CWJ and P Newbold (1974). Spurious regressions in econometric model specification. *Journal of Econometrics* **2**, 111–120.

Hamilton, JD (1994). *Time Series Analysis*. Princeton: Princeton University Press.

Harris, R (1995). *Using Cointegration Analysis in Econometric Modeling*. London: Prentice Hall.

Holden, D and R Perman (1994). Unit roots and cointegration for the economists. In *Cointegration for the Applied Economist*, BB Rao (Ed.), UK: St. Martin's Press, 47–112.

Johansen, S (1995). *Likelihood-Based Inference in Cointegrating Vector Autoregressive Models*. New York: Oxford University Press.

Johansen, S and K Juselius (1990). Maximum likelihood estimation and inference on cointegration with application to the demand for money. *Oxford Bulletin of Economics and Statistics* **52**, 169–210.

Kashyap, KA and CJ Stein (1994). Monetary policy and bank lending. *NBER Studies in Business Cycles* **29**, 221–256.

Kashyap, KA and CJ Stein (2000). What do a million observations on banks say about the transmission of monetary policy? *The American Economic Review* **90**(3), 407–428.

Kashyap, KA, AO Lamont and CJ Stein (1994). Credit conditions and the cyclical behavior of inventories. *The Quarterly Journal of Economics* **109**(3), 565–592.

King, GR and R Levine (1993). Finance, entrepreneurship and growth: theory and evidence. *Journal of Monetary Economics* **32**, 1–30.

Lazaridis, A (2007). A note regarding the condition number: the case of spurious and latent multicollinearity. *Quality & Quantity* **41**(1), 123–135.

Levine, R and S Zervos (1998). Stock markets, banks, and economic growth. *American Economic Review* **88**, 537–558.

Lown, C and DP Morgan (2006). The credit cycle and the business cycle: new findings using the loan officer opinion survey. *Journal of Money, Credit, and Banking* **38**(6), 1575–1597.

MacKinnon, J (1991). Critical values for co-integration tests. In *Long-Run Economic Relationships*, RF Engle and CWJ Granger (eds.), pp. 267–276. Oxford: Oxford University Press.

Mckinnon, RL (1973). *Money and Capital in Economic Development*. Washington, DC: Brookings Institution.

Meltzer, HA (1995). Monetary, credit and (other) transmission processes: a monetarist perspective. *The Journal of Economic Perspectives* **9**(4), 49–72.

Mishkin, SF (1995). Symposium on the monetary transmission mechanism. *The Journal of Economic Perspectives* **9**(4), 3–10.

Mouza, AM and D Paschaloudis (2007). Advertisement and sales: a contemporary cointegration analysis. *Journal of Academy of Business and Economics* **7**(2), 83–95.

Murdock, K and J Stiglitz (1997). Financial restraint: towards a new paradigm. *Journal of Monetary Economics* **35**, 275–301.

Rousseau, PL and P Wachtel (1998). Financial intermediation and economic performance: historical evidence from five industrialized countries. *Journal of Money, Credit and Banking* **30**, 657–678.

Rousseau, PL and P Wachtel (2000). Equity markets and growth: cross-country evidence on timing and outcomes, 1980–1995. *Journal of Banking and Finance* **24**, 1933–1957.

Shaw, ES (1973). *Financial Deepening in Economic Development*. New York: Oxford University Press.

Sims, CA (1972). Money, income, and causality. *The American Economic Review* **62**(4), 540–552.

Sims, CA (1980a). Comparison of interwar and postwar business cycles. *The American Economic Review* **70**, 250–277.

Sims, CA (1980b). Macroeconomics and reality. *Econometrica* **48**, 1–48.

Stiglitz, EJ (1992). Capital markets and economic fluctuations in capitalist economies. *European Economic Review* **36**, 269–306.

Stock, JM and MW Watson (1988). Testing for common trends. *Journal of the American Statistical Association* **83**, 1097–1107.

Stock, J and MW Watson (1989). Interpreting the evidence on money-output causality. *Journal of Econometrics*, January, 161–181.

Swanson, N (1998). Money and output viewed through rolling window. *Journal of Monetary Economics* **41**, 455–473.

Tobin, J (1970). Money and income: post hoc ergo propter hoc? *Quarterly Journal of Economics* **84**, 301–329.

Trautwein, M (2000). The credit view, old and new. *Journal of Economic Surveys* **14**(2) 155–189.

INDEX